How We Govern Our Minds Through Others

How We Govern Our Minds Through Others

Epistemic Autonomy Beyond the Myth of Independence

J. Adam Carter and Neil Levy

The MIT Press
Cambridge, Massachusetts
London, England

The MIT Press
Massachusetts Institute of Technology
77 Massachusetts Avenue
Cambridge, MA 02139
mitpress.mit.edu

The MIT Press would like to thank the anonymous peer reviewers who provided comments on drafts of this book. The generous work of academic experts is essential for establishing the authority and quality of our publications. We acknowledge with gratitude the contributions of these otherwise uncredited readers.

This book was set in Stone Serif and Stone Sans by Westchester Publishing Services. Printed and bound in the United States of America.

Library of Congress Cataloging-in-Publication Data

Names: Carter, J. Adam, 1980- author | Levy, Neil, 1967- author
Title: How we govern our minds through others : epistemic autonomy
  beyond the myth of independence / J. Adam Carter and Neil Levy.
Description: Cambridge, Massachusetts : The MIT Press, [2026] |
  Includes bibliographical references and index.
Identifiers: LCCN 2026006611 (print) | LCCN 2026006612 (ebook) |
  ISBN 9780262056458 paperback | ISBN 9780262056465 pdf |
  ISBN 9780262056472 epub
Subjects: LCSH: Autonomy (Philosophy) | Knowledge, Theory of |
  Social epistemology
Classification: LCC B808.67 .C37 2026 (print) | LCC B808.67 (ebook)
LC record available at https://lccn.loc.gov/2026006611
LC ebook record available at https://lccn.loc.gov/2026006612

EU Authorised Representative: Easy Access System Europe, Mustamäe tee 50, 10621 Tallinn, Estonia | Email: gpsr.requests@easproject.com

Contents

Preface vii

1 **Living with Epistemic Dependence** 1
2 **Hijacked Attention** 21
3 **Group Doxastic Autonomy** 37
4 **Epistemic Autonomy in the Age of AI** 65
5 **Negative Autonomy, Dependence, and Epistemic Axiology** 85
6 **Nudging Inquiry: Erotetic Attitudes and Intellectual Autonomy** 103
7 **Epistemic Autonomy in a Loop-Ridden Landscape: Navigating Social-Epistemic Dependencies** 119

Notes 137
References 155
Index 183

Preface

0.1 Epistemic Dependence in (Cartesian) Context

If philosophy begins in wonder, as Aristotle suggested, epistemology began in doubt. Cartesian doubt, generalized doubt, might represent the founding moment of modern epistemology, and much of its subsequent history is concerned with rebutting the skeptical challenge. How can we know anything if we can't rule out the possibility that we're dreaming or being deceived by a powerful and malign being?

The idea that modern philosophy begins with Descartes's skeptical challenge is a familiar one. What's less familiar is that Descartes himself actually begins not with the worry that he might be deceived but instead with the idea that will play the starring role in this book: *epistemic dependence*.

On the fateful day in 1619 when Descartes, left alone with his thoughts in a stove-heated room, resolved to rebuild knowledge from the ground up, the evil genius was not his first concern. Rather, he began with the recognition—or what he took to be recognition, at any rate—that things "produced by the hands of several master craftsmen are often less perfect than those on which only one person has worked" (Descartes 2006, 12). This was true, Descartes thought, when it came to buildings and to towns and to laws: When they evolved over time, and therefore represented the work of many different individuals, they were imperfect because they were not "directed to a single end" (13). If this is a concern when it comes to urban planning, how much more is it a worry when it comes to knowledge?

Descartes was well educated in mathematics, physics, and the law, but could he rely on what he read in the libraries of Poitiers or Paris? The education he had received, no less than the streets he walked, represented the work of many hands. "Book-learning," too, is unreliable, "composed as it is of the opinions of many different people" (2006, 13). Except for the mathematical proofs he was able to verify for himself, Descartes resolved to cast away everything he had learned as the merest opinion of others.

It's no accident that Descartes began not with doubt but rather with epistemic dependence. Individualism and skepticism are inextricable: If you believe that knowledge is secure only when it is always and everywhere an unaided individual achievement, you must doubt most of what you might have thought you knew. Epistemic individualism forces the Cartesian project on us. If that's our starting point, we have to raze belief to the ground and begin anew. Conversely, if we are finally to put a generalized skeptical challenge to rest, we must first banish epistemic individualism.

Contemporary epistemology is much less concerned with the skeptical challenge than it was in the past. Most of us have recognized that indubitability is neither to be had nor needed. We can know many things, even if we can't banish the skeptical demons. One of Descartes's challenges has not so much been addressed as lost much of its interest. But the second Cartesian challenge is more pressing today than ever before. We live in a world of expert knowledge. In every arena of our lives, we are dependent on epistemic advice and expert guidance—on knowledge claims we are unable to verify for ourselves. Moreover, and much worse, concerns about the reliability of this expert knowledge are no mere thought experiment. There are many *actual* cases in which expert guidance has been shown to be wrong or even corrupt.

We might think here of the opioid epidemic. Since the 1990s, more than eight hundred thousand Americans have died from opioid overdoses (Centers for Disease Control and Prevention 2023). Although a variety of drugs, legal and illegal, underlie these deaths, the epidemic began with prescription drugs, and directly and indirectly (through people who began with prescribed opioids before switching to illegal drugs), they remain big killers. Unreliable advice from experts played a significant role in the opioid epidemic. Doctors prescribed them to their patients in doses and over timescales that we now know to be inappropriate. Many of these doctors acted in good faith. They relied on advice from experts and on guidance from the US Food and Drug Administration (FDA), which has the job of approving and labeling medications. We now know that many of the experts who advised liberal use of opioids as analgesics were co-opted by the pharmaceutical companies (Humphreys et al. 2022), and that the FDA failed in its duty to protect the public (Kolodny 2020). FDA failures, in turn, reflected under-funding and under-resourcing, with conflicts of interest likely playing a role as well. We don't need to imagine cases in which expert guidance is unreliable: They are all too real.

Although we do not share Descartes's conviction that the products of many hands are inevitably inferior to the products of a single designer—this is, after all, a coauthored book—it's much harder to shake off worries about epistemic dependence than it is to dismiss brain-in-a-vat skepticism. Epistemic dependence is both indispensable and often highly beneficial: We know far more because we depend on and collaborate with one another than we could possibly know by ourselves. Perhaps—*perhaps*—godlike beings could dispense with such dependence and be none the worse epistemically for it. But that's not an option for us. We must live *with* epistemic dependence. We must find a way to live in a world in which much of what we need to know we cannot know on our own, but in which we also know ourselves to be subject to manipulation by those who would exploit us.

This book explores some of the many issues our epistemic dependence raises. We take for granted here that we cannot dispense with epistemic dependence. Expertise is hard won, and each of us can hope to possess it only in a narrow sphere. Although there are various heuristics we can use to identify reliable experts (Goldman 2001; E. Anderson 2011; Grundmann 2025), these heuristics only get us so far: They allow us to filter out only the more unsophisticated frauds. Cues to expertise can be mimicked, and canny fraudsters take advantage of that fact (Guerrero 2017; Levy 2018a). Moreover, as the opioid epidemic illustrates all too starkly, genuine expertise can be suborned. We do not believe that the prospects for verifying expertise for ourselves are good (Levy 2022b). Instead, we must live with the uncomfortable knowledge that our dependence is ineliminable. The above guiding (and related) issues emerge and reappear (and often in conversation with each other) throughout the book's chapters, which we detail in what follows.

0.2 Overview of Chapters

How We Govern Our Minds Through Others is organized around seven substantial chapters, each of which approaches the puzzle of epistemic autonomy—and philosophical questions in its constellation—from a different angle.

The central themes explored in a sustained way in the book include epistemic autonomy in connection with hijacked attention, intellectual collaboration with others in group settings, artificial intelligence (AI), firsthand insight, nudging and questioning, epistemic feedback loops, and adaptive control. What emerges is a reconceptualization of epistemic autonomy of the sort worth valuing as deeply social and deeply scaffolded.

Chapter 1, "Living with Epistemic Dependence," argues that agents like us are epistemically vulnerable in virtue of our dependence on one another. The chapter begins with good news: Despite widely circulated claims to the contrary, there's little evidence of widespread gullibility. We are not gullible because we possess sophisticated mechanisms of epistemic vigilance—mechanisms that are designed (inter alia) to assess messages for credibility. We possess these mechanisms because we are at one and the same time not only deeply dependent on one another but also in competition with one another. We must learn from one another, but we must also protect ourselves from exploitation. We argue that although these mechanisms do an excellent job at filtering out claims that are bizarre by the standards of our society, they nevertheless leave us vulnerable to misinformation when its source passes tests for credibility. Living with epistemic dependence requires some way of living with the fact that we often lack any independent way of assessing the testimony we need.

Chapter 2, "Hijacked Attention," highlights a central limitation of the mechanisms of epistemic vigilance discussed in the first chapter. These mechanisms are excellent protection against accepting beliefs that strongly conflict with those that our broader culture takes seriously. They ensure we are unlikely to accept bizarre misinformation. In the first chapter, we noted that they nevertheless leave us vulnerable to certain kinds of misinformation. In this chapter, we explore harms that arise from the more bizarre misinformation that epistemic vigilance filters out as a candidate for belief. These mechanisms offer us no protection against attentional capture. We argue that attentional capture threatens to thwart or hijack debates and that we have few defenses against it.

Chapter 3, "Group Doxastic Autonomy," shifts perspective from individuals to groups by asking whether group beliefs can count as autonomous or heteronomous. Although we develop and defend a non-summativist account of group beliefs (i.e., an account on which group beliefs can dissociate from the beliefs of individual group members), our primary aim is to set out the options, focusing in particular on whether group autonomy in belief is itself summativist (i.e., on its relationship to individual level autonomy in belief). We argue that there's a case for thinking that group autonomy can dissociate from individual-level autonomy on every available account, which suggests that group-level autonomy should be conceived of non-summatively.

In chapter 4, "Epistemic Autonomy in the Age of AI," we consider the extent to which AI systems—we focus on large language models (LLMs) such as ChatGPT—might threaten or, on the contrary, support epistemic autonomy. We begin with ways in which LLMs might scaffold the development of the capacities needed for epistemic autonomy, especially an appropriate degree of self-trust. We argue that peer disagreement (with a human being) is often uncomfortable because it threatens our self-conception as an epistemic equal. This anxiety, in turn, may hinder the development of the self-trust needed for epistemic autonomy. But the disposition to be appropriately firm in peer disagreement is essential to epistemic autonomy. We suggest that this disposition might be scaffolded by "disagreement" with an LLM. In this way, LLMs might support the development of the capacity for epistemic autonomy. But they might also have the potential to undermine it. LLMs seem to fall somewhere between being mere cognitive tools and genuine collaborators in epistemic enterprises. This ambiguity prevents agents from taking responsibility for the results of their interaction with LLMs, thereby preventing satisfaction of the ownership condition on autonomy. We finish by noting that collaborative knowledge production might already threaten epistemic autonomy in an analogous way, and questioning whether this should be concerning.

Chapter 5, "Negative Autonomy, Dependence, and Epistemic Axiology," interrogates the value of intellectual independence by challenging the negative epistemic autonomy (NEA) thesis—the idea that, all else being equal, knowledge gained through independent inquiry is more epistemically valuable. Although acknowledging the value of goods such as "understanding-why" and "strong cognitive achievement," as highlighted by thinkers such as Ernest Sosa and Duncan Pritchard, the chapter argues against conflating these with the first-order epistemic independence that accompanies solitary inquiry. The view arrived at is that this independence, in the course of inquiry, holds, at best, derivative, instrumental value as a shield that enables, rather than enhances, achievement in attention-hijacking environments.

Chapter 6, "Nudging Inquiry: Erotetic Attitudes and Intellectual Autonomy," shifts focus from beliefs to questions. Whereas literature on epistemic nudging has primarily focused on influencing belief formation (doxastic nudging), the zetetic turn in epistemology highlights the importance of interrogative attitudes in inquiry. This chapter explores the overlooked space of nudging interrogative attitudes, such as wonder and curiosity, and their relation to the uptake of misinformation. It identifies three distinct species of erotetic (interrogative-attitude directed) nudging: (a) generative, (b) presuppositional, and (c) partition-framing. It argues that erotetic nudging often promotes erotetic autonomy by providing reasons to inquire in epistemically beneficial ways, while also considering the boundaries between nudging and manipulation. This framework demonstrates the potential of interrogative nudging as a distinctive social-epistemic strategy for combating misinformation at its source in problematic questions.

Chapter 7, "Epistemic Autonomy in a Loop-Ridden Landscape: Navigating Social-Epistemic Dependencies," investigates how agents, as they increasingly rely on technology and others for cognitive tasks, risk getting caught in feedback loops, some epistemically beneficial and others harmful. This chapter argues that the pervasiveness of such loops in our social-epistemic environment has important implications for epistemic autonomy. The more we depend on epistemic scaffolding, the more crucial it becomes to cultivate capacities to avoid getting trapped in performative loops that misalign with our epistemic interests. This challenge is conceptualized as a special case of an adaptive control problem. By developing strategies to minimize the risk of being pulled into misaligned loops while intentionally participating in beneficial ones, agents are better equipped to navigate this loop-ridden landscape more effectively, with the caveat that not all such loops are escapable, given our finite capacities.

On the title page, the author order is alphabetical; we each contributed equally to this book.

The book does not purport to address definitively, or even partially, the full range of ways in which our epistemic dependence might give rise to the sort of Cartesian anxiety we identified above. Our aim is much less grand. We aim to offer diagnoses of some of the many ways in which our epistemic autonomy, as individuals and as members of groups, can be threatened, as well as to offer some ways in which we can assert a measure of epistemic autonomy in the face of these challenges. Nothing we say is definitive: Some is explicitly offered with the aim of sketching options. If we are to become epistemically autonomous, we might say, we cannot do it alone. Our dependence extends even to our autonomy. We hope to contribute to a conversation in which, together, we can better manage our epistemic commons.

We are grateful to the Arts and Humanities Research Council for funding the research that led to this book via the AHRC project "Expanding Autonomy: Scaffolded, Embedded and Distributed" (grant ref. AH/W005077/1) at the Universities of Oxford and Glasgow. That grant enabled us to appoint two very able post-docs, Oscar Piedrahita and Matthew Vermaire, who provided invaluable feedback on some of the material presented here, as did Emma Gordon. Christoph Kelp, Mona Simion, Jonathan Pugh, Cristina Voinea all participated in a workshop on the book that led to valuable insights.

Carter's work on this monograph is also supported by the European Research Council grant number 101199255 (KNOW-HOW).

Chapter 2 contains material from Levy, Neil, "Capturing Our Attention," *Australasian Journal of Philosophy,* forthcoming.

1 Living with Epistemic Dependence

The word "autonomy," like many words expressing philosophically central concepts, derives from Greek: *auto*, meaning "self," and *nomos*, meaning "laws." To be autonomous is to give oneself the laws one abides by—to rule oneself.[1] Although etymology does not determine meaning, contemporary uses of "autonomy" take its literal meaning seriously. In political thought, a nation is autonomous when it rules itself, and in moral thought, autonomy is understood to consist in the capacity to decide for oneself. Accordingly, an autonomous attitude or thought is one the agent has decided for themselves or that is an expression of oneself.[2]

This capacity to decide for ourselves is something we value. Autonomy is closely connected to self-respect: The person who lacks it, and for whom others must decide, may not be able to respect themselves as an equal among others. But any account of autonomy must grapple with our actual dependence on one another. Feminist theorists have long been critical of what Marilyn Friedman (2003) once derided as the Marlboro Man model of autonomy, on which autonomy consists in moral and practical self-sufficiency. Our relations to one another are morally central to who we are. They are neither optional extras nor regrettable embarrassments. The truly self-governing self, a self extracted from the network of constitutive relationships, obligations, and dependencies, neither exists nor would be valuable if it did.

The Marlboro Man model obscures, moreover, the work that others—disproportionately women—do in bringing individuals to the point where they can stand on their own two feet, to the extent they ever can. Behind every man declaring he is self-made are a number of people, typically women, on whom he has been and still is dependent. The Marlboro Man model hides subordination. It is also ethnocentric. Work on informed consent, in particular, has emphasized how many cultures value family members deciding together rather than individuals deciding for themselves (Fan 2015). The Marlboro Man model reflects the psychology of WEIRD (Western, educated, industrialized, rich, and democratic) nations (Henrich et al. 2010; Henrich 2020), and this psychology has no unique claim to moral correctness.

Nowhere are the difficulties with understanding autonomy as self-rule more apparent than in epistemology. Prima facie, epistemic autonomy, self-rule in the domain of knowledge and belief, must be valuable if autonomy is valuable at all: Beliefs are obviously central to decision-making, and we govern ourselves largely by making decisions. Although the relationship between epistemic autonomy and personal autonomy is not one-to-one—we may be able to act autonomously on the basis of beliefs that are not themselves autonomous—there's a strong case for thinking that our actions can express our values only if our beliefs are very largely autonomous. But as we'll see, it's also hard to see how we can be epistemically autonomous.

"Self-Reliance," Emerson's (1841) famous essay against conformist tendencies, toyed in places with the idea that, epistemically speaking, we should be entirely self-reliant. This was an ideal he presented as a contrast to what is nowadays called a "sheep" mentality. As Emerson put it, "It is easy in the world to live after the world's opinion. It is easy in solitude to live after our own. But the great man is he who in the midst of the crowd keeps with perfect sweetness the independence of solitude" (55). Even if we agree with Emerson that there is a disvalue to living one's life as a thoroughly uncritical "sheep," it remains true that keeping with, and being guided by, just the "sweet solitude" of one's own opinions—ignoring or otherwise bracketing good information in our environments simply because it is others and not ourselves who have worked for it—is nonetheless epistemically disastrous. After all, as believers, we are deeply and necessarily dependent on others for an enormous number and range of our beliefs. Testimony, understood broadly, is the source of *most* of our beliefs (Roeber 2024, 21). Beyond the narrow range of what we can hear, see, and feel for ourselves and what we can deduce for ourselves, we are utterly dependent on others. We know about the past, including our *own* past, only because others have told us. We know about other places only because others have told us (even if we've visited, we saw little of it). We know about our own country, even our village, only because others have told us. Even for those closest to us, we know about their lives only because they share with us. Subtract from us our beliefs arrived at via deference, and we know very little.

Whatever lesson there is to take from Emerson's criticism of intellectual sheep, it shouldn't be taken as a call for epistemic *independence*. Instead, it would need to be reconciled with an undeniable truth: Our dependence on others runs deeper than Emerson imagined—and grows deeper still in our Digital Age. A 2018 Ofcom report found that UK residents spend an average of twenty-four hours online each week, and that they check their phones approximately every twelve minutes during waking hours—equating to roughly eighty checks per day (assuming sixteen waking hours). What we are inundating ourselves with on each check is, in one way or another, information curated and framed by others.

A challenge similar to Emerson's appears in Kant's ([1784] 1991) influential account of enlightenment. At first glance, Kant seems to champion a kind of epistemic Marlboro Man when he contrasts Enlightenment ideals with what he calls intellectual "immaturity." For Kant, immaturity stems from "lack of determination and courage to use one's intelligence without being guided by another." His rallying cry *Sapere Aude!*—"Have the courage to use your own intelligence!"—is for him the motto of the Enlightenment ([1784] 1991, § 1). Let's grant that there is some truth in the motto.

But what does it mean to "use one's own intelligence" in the way that avoids intellectual immaturity? Although Kant himself embraced the Marlboro Man model, maintaining that intellectual maturity consisted in the ability "to use one's own understanding without the guidance of others," using "one's own intelligence" need not—and cannot—exclude appropriate reliance on others. We use our own intelligence when we critically depend on one another. Intellectual immaturity, on this conception, consists in *bypassing* this reasoning process entirely—simply accepting others' conclusions about what we should intend, as when we uncritically follow orders. Under such conditions, we are not the author of any laws we give ourselves.

The mature alternative to Kantian immaturity does not consist in solitary reasoning, or in reasoning that accepts only premises one has established for oneself. Suppose I desire to visit London (having read about interesting shows there this weekend), I believe the 7:30 a.m. train from Edinburgh Waverley will get me there (based on a colleague's advice), and I therefore form the intention to take that train. Here, I reason through the practical syllogism myself—while depending entirely on others for the support

I have for my premises. My desire stems from others' reports about London shows; my belief about the train comes from testimony. Without such dependence, I likely couldn't reach London at all if I tried, and wouldn't even desire to, at least on that occasion for that purpose, in the first place.

It is tempting to think that autonomy requires "using one's own reason" via explicit deliberation—even if the support we have for our premises (in either practical or theoretical reasoning) is highly dependent on others. But it may be (and this is a point we'll return to in various places) that even *this* is a step too far in the direction of the Marlboro Man. For a start, policies to forgo deliberation may be arrived at rationally. I might know, for example, that when it comes to medical decisions about COVID, I am not great at evaluating evidence, and I might know that my doctor is fantastic. Am I not following Kant and using my own reasoning to arrive at a policy that *outsources* on such matters both what I should believe and what I should intend (even if I don't outsource to my doctor what music to listen to or where to buy a car)?

At any rate, closer inspection suggests a charitable reading of the value of "avoiding immaturity" is going to need to make peace with the fact that our epistemic dependence is deep. This dependence is perhaps most complete in some of the domains that matter most deeply to us, such as morality and politics, and also when it comes to making personal decisions. Agents know about the policies of political candidates via testimony rather than by any kind of direct observation. They know about the implications of these policies, to the extent they do—Will they promote economic growth? Limit freedoms? Impinge on rights?—only because others analyze them for us. We form beliefs about deeply personal decisions, from investing in the stock market to taking vaccines (or refusing them) and even about such matters as whether to have children, partly on the basis of testimony (What was it like for our friends? How much of a financial and emotional struggle was it—and how much joy did it bring them?). How can we be autonomous in the face of this deep dependence?

We don't pretend definitively to answer that question in this book. Our aim is much less ambitious: to identify and to explore different dimensions of the question and to get a better sense of the options and their costs. That said, we will not hesitate to offer opinions on which options are best, when we have them, and to defend these opinions. Our first order of business is to get clearer on the nature of the challenge to epistemic autonomy.

In the rest of this introductory chapter, we will show that one influential way of characterizing a central dimension of the problem is mistaken. In one respect, we will argue, the news is much better than many people think. In the subsequent section, however, we will argue that, in another respect, the news is worse than is commonly thought. The good news: We are—contrary to a familiar line of thinking in social epistemology (Kahan 2013)—neither gullible nor irrational. The bad news: Epistemic dependence ineliminably goes much deeper than many theorists recognize, even more so than we've suggested so far. We will suggest that the very mechanisms that protect us from excessive credulity under many conditions entail a deep and ineliminable epistemic vulnerability.

1.1 The Good News!

Autonomy is closely linked to rationality. Raz (1986), for example, takes it that to be the author of one's life, one must possess significant mental capacities—including a minimum rationality—to form complex intentions and plan their execution. As Raz puts it, "these include minimum rationality, the ability to comprehend the means required to realize [one's] goals, [and] the mental faculties necessary to plan actions, etc." (1986, 373). Without such rational capacities, a person could not meaningfully choose or plan; autonomy would be impossible.

In this light, many people see our apparent flaws as reasoners to be a major threat to, or at any rate a limitation on, autonomy. As Jonathan Pugh (2020) puts the point, "theoretical irrationality" defeats autonomy in the sense that autonomy presumes at least basic understanding and coherence, so decisions based solely on error or delusion are not autonomous decisions. This lines up with the idea that someone who acts impulsively, under mania or from deeply irrational prejudice and so forth, is not really "self-governing."

In this section, we argue that the threat from irrationality to epistemic autonomy has been exaggerated by many commentators. We will focus on one particular form this threat is widely supposed to take (although we think the lesson can be generalized to other forms of alleged irrationality, such as our vulnerability to cognitive biases of various sorts). The threat we are concerned with arises from misinformation.

The past decade has indeed been gripped by a near panic about misinformation, especially in the guise of "fake news" (Gelfert 2018; Balakrishnan et al. 2022) and, increasingly, "deep fake" images, audio, and video (Rini 2020; cf. Habgood-Coote 2023). What else, many commentators ask, could explain the series of catastrophically bad choices that democratic nations have made since 2016 (the year of the Brexit referendum)? Fake news proliferates across social media, and people seem increasingly to embrace bizarre conspiracy theories. Ordinary people are not merely dependent on testimony, they're also downright gullible. They're eager to accept anything that flatters their sense of self and that panders to their prejudices, no matter how bizarre the claim might be—hence, for example, the rise of QAnon, an amorphous conspiracy theory, according to which deep-state actors and senior politicians operate a child sex-trafficking ring and plot against the interests of the American people (Rose 2020; Bleakley 2023).[3]

These conspiracy theories are apparently embraced by huge numbers of people.[4] At its height, for example, polling found 50 percent of Americans reporting believing in or being unsure about the truth of QAnon (Ipsos 2020). Commenting on these figures, one pollster concluded that his fellow Americans had gone "bonkers" (Rose 2020). Epistemologists and social scientists have shared his concern. Some have gone so far as to speak of a misinformation *crisis* (Houlden et al. 2021; Allchin 2022; Germani et al. 2022; O'Connor and Weatherall 2019).

Underlying these concerns is a particular conception of the harms of misinformation. On the most common view, misinformation is harmful when and because it causes *false beliefs*.[5] For Flynn et al., for example, misinformation is harmful because it leads to *misperceptions*: "factual beliefs that are false or contradict the best available evidence in the public domain" (2017, 128). Such misperceptions are harmful, in turn, because they distort debate: We cannot agree on a solution to a problem if we cannot agree on the nature of the problem (e.g., whether global temperatures are rising). Misperceptions also cause harms more directly—for example, by leading people to avoid vaccines for themselves or for their children. As we've already noted, moreover, misperceptions threaten autonomy: They arise from irrational processes, and they undermine our capacity to express our values and to pursue our goals.

We have no doubt whatsoever that misinformation can and does mislead. Given our pervasive reliance on testimony, false testimony, or even accurate testimony that deceives through misleading conversational implicatures (Simion 2021a), can and does lead to ignorance under a variety of circumstances. But the evidence strongly suggests that very many of those who report believing misinformation or who share it, apparently seriously, don't really believe it.

People may report believing something they don't really believe either because they're insincere or because they're mistaken about their own beliefs. There is plentiful evidence that people who report believing in QAnon, or who share QAnon supportive posts on social media, often fall into one or the other of these two categories.

People report beliefs insincerely for at least two reasons: to express their support for a view, a group, or a person, or to troll the people putting the questions. There's empirical evidence for both mechanisms.

Expressive responding—reporting a belief one does not hold in order to express a conative attitude—is now the subject of a fairly large literature (Berinsky 2018; Hannon 2021). Democrats and Republicans report divergent beliefs about the state of the economy, but these reports may express their attitudes to the party that happens to be power at the time rather than their veridical perceptions, which we may conclude from the fact that economic behavior seems better predicted by the actual state of the economy than their reported beliefs (Bullock and Lenz 2019) and that small incentives for accuracy narrow the partisan gap (Prior et al. 2015; Bullock and Lenz 2019).

Admittedly, some of these data are open to an alternative explanation. Perhaps people report the belief they believe *the experimenters* hold rather than their own genuine belief in order to earn money for an accurate response. But some experimental evidence of expressive responding is compelling. In the immediate aftermath of the controversy surrounding the (false) claim by Trump's then–press secretary that his inauguration had been the largest on record, Brian Schaffner and Samantha Luks (2018) gave their respondents unlabeled photos of the Trump and Obama inaugurations, and simply asked them which photo had more people (figure 0.1). They expected that some respondents would identify the photographs and take the opportunity to express their support for Trump by expressive responding. Indeed, a significant minority of Trump voters (but not Clinton voters or independents) identified the photo of the Trump event as depicting the larger crowd (see Ross and Levy [2023] for replication). It's not really plausible that these participants were reporting a genuine belief: The disparity between the crowds depicted was too stark. Expressive responding is a far more plausible explanation.

Figure 0.1
Photos of the Trump inauguration in 2017, left, and the Obama inauguration crowd in 2009, right, used as stimuli in Schaffner and Luks (2018).

Expressive responding has received a significant amount of attention in psychology (and some within philosophy; Hannon 2021). Another motivation for insincere belief report has received much less attention: sheer trolling. Some people openly report that they create or share fake news "for the lulz" (Kunzru 2020), but there's been little systematic research on this sort of behavior (see Lopez and Hillygus [2018] for a rare exception). On some topics, trolling seems to be a powerful force. Consider another first Trump-term topic: the use of household bleach to fight COVID infections. After Trump publicly speculated on its value, there were numerous media reports that credulous Trump supporters were ingesting or injecting bleach (Joseph 2020; Reuters 2020; Reimann 2020). Going beyond anecdotal reports, the Centers for Disease Control and Prevention (CDC) conducted an Internet panel survey of just over five hundred US adults and found that 4 percent reported ingesting bleach (Gharpure et al. 2020). If the population surveyed was representative, that's millions of people consuming bleach.

But are the reports reliable? Litman et al. (2023) conducted two follow-up surveys, each with a larger participant pool than the CDC study. They replicated the result with regard to bleach, finding, in addition, that other participants reported drinking or gargling other household cleaners. But the results were very different when they removed "problematic responders," such as the 5.8 percent of respondents from the first survey who reported having suffered a fatal heart attack, who failed attention checks, or who reported knowing what the made-up word "wuttlet" means. This procedure reduced the pool of those reporting treating COVID with household cleaners to just twelve individuals. Nine of these twelve reported accidentally choosing "yes" in response to being asked whether they'd ingested cleaning products. A further two reported they'd drunk these products by accident. That left a single individual who claimed intentionally to have ingested cleaning products to fight COVID. This same individual reported their weight as 1,900 pounds. In the light of this evidence, we have every reason to think that approximately zero people deliberately drank bleach to combat COVID, media reports notwithstanding.

Trolling and expressive responding are two distinct routes to insincere belief reports. Sincere but mistaken belief reports are also produced in two distinct ways. One route is via the *construction* of reports. It's long been recognized that a great many survey responses are constructed on the spot. When asked a question that one has never considered before in any depth, one *constructs* an attitude by sampling the considerations that come to mind (Zaller and Feldman 1992).

Reporting an attitude about something we've never thought about—even if we are aware we've never thought about that topic, which is not always the case—is perfectly appropriate. We often, perhaps typically, probe our own attitudes not by recalling what we have thought on previous occasions but instead by considering what our attitude *should* be, given the evidence. We look to the world, that is, and not to our own mental states. Asked "Do you think Donald Trump will declare martial law?," agents probably don't recall previous beliefs; more likely, they think about the world and about Donald Trump. Belief reports such as these are, as Evans (1982) put it, transparent: We report our beliefs not by looking within but rather by looking at the world our beliefs aim to represent.

We also report beliefs we've never articulated by thinking about what our beliefs entail. Many of our dispositional beliefs are reported via this sort of route. Do you believe there's a school of tuna living on the asteroid known as Bennu (a small near-Earth asteroid)? Before writing this sentence, neither of us had considered the question, but we're confident in the answer: It's entailed by our settled beliefs about asteroids and fish. Given how normal and appropriate it is to report beliefs on topics we've never considered, it's not surprising that survey respondents and experimental subjects report beliefs on topics they've never previously thought about.

In most circumstances, the construction of attitudes via one route or another probably leads to the reporting of genuine beliefs. The attitude reported is a stable state because it's entailed by beliefs that are themselves stable. But often, constructed belief reports do not index stable states, either because the attitudes on the basis of which we infer them are not themselves stable or because these attitudes have an imprecise content and therefore do not entail unique beliefs. Someone's response to the question of whether Donald Trump will declare martial law might not report a stable state because they sample only a small range of their own prior attitudes in constructing it. This sort of unstable response is particularly likely when people are asked questions about which they have little information—which is often the case when they're asked about conspiracy theories or political attitudes. But it also occurs when our attitudes are hard to cash out with precision. Consider, for example, someone's sincere belief that everyone should be treated equally. Most of us have only vague answers to questions such as "Equality of what?" and a correspondingly vague sense of in what respects people should be treated equally (Schwitzgebel 2009). Because we only have vague answers to such questions, belief reports that depend on them may have contents that shift over time (Levy 2018b).

These sorts of reflections explain much of the instability of belief reports seen in the empirical literature, we believe. When people report knowledge, they report entailments from their stable beliefs or they retrieve a state from semantic memory, and the report is stable over time, but when they are asked for their attitude about some political topic on which they haven't previously reflected, the considerations that come to mind may be unrepresentative or outweighed by others they fail to consider, and the resulting report will not be stable over time (M. H. Graham 2023). These beliefs are, in Lackey's (2021) terminology, "judgmentally fragile": Were the person to deliberate on a different occasion, they might easily report a different belief.

We think we should not regard the states reported as beliefs. They do not cause behavior beyond the assertion: When stakes rise, people are apt to sample a broader range of considerations. This may explain why conspiracy theorists act consistently with their reported beliefs relatively rarely (Mercier 2020).[6]

The second route to sincere but mistaken belief reports is via first-person error, when people mistake other sorts of states for beliefs. The mind is opaque (Carruthers 2013; Schwitzgebel 2008), and identifying our own beliefs and distinguishing them from other attitudes is not always easy. When an attitude's content is imprecise, and therefore allows us all sorts of wiggle room with regard to behavior and inference, we may mistake a vividly held imagining for a belief. We think this is true with regard to many of the more bizarre beliefs people profess to hold: flat-earth beliefs, beliefs in wild conspiracy theories, and so on (Levy 2022a). That they are some sort of imaginings—*shmeliefs*, not beliefs (Levy 2024a)—is indicated not only by their failure to govern behavior broadly and their lack of responsiveness to evidence but also and especially by the "evidence" that those who report them are prone to cite. When people apparently seriously point out that "delta omicron is an anagram of media control" in support of their belief that the COVID pandemic was preplanned, they reveal that, at least at some level, they are aware that they're not really responding to evidence or reporting beliefs.

We do not suggest that no one seriously believes the planet Earth is flat, or that Democrats are sacrificing children to Satan, or that the dogs and cats of Springfield, Ohio, are being systematically eaten by immigrants. On all these topics, there likely are some true believers, as well as shmelievers and sheer trolls. One and the same person may start as a troll and become a true believer, perhaps with an intermediary shmelief stage, and some people may oscillate between shmelief and belief. As a general rule, the more bizarre the belief, the higher the proportion of shmelievers among those who sincerely assert it, we suspect.

The evidence that people are much less gullible than is commonly thought seems hard to square with a large literature that seems to show that we are extremely bad at detecting deception (e.g., Hartwig and Bond 2011; Bond Jr. and DePaulo 2006). Even those with a professional interest in identifying lies, such as police officers, detectives, judges, and psychologists, perform at near-chance levels (Aamodt and Custer 2006). In the light of this evidence, one might expect people to be easily taken in by con artists of all sorts.

If we really were pervasively vulnerable to deception, we would not see widespread gullibility. Rather, we would see the breakdown of communication altogether. Cooperative arrangements would unravel in the way that has become familiar in evolutionary models of interaction strategies: Those who do not trust others would outcompete more trusting conspecifics, and we would no longer be a social species. Obviously, though, we are a social species, and we routinely engage in communicative exchanges. We must have some way of distinguishing more and less reliable communicative signals or ensuring that we are much less exposed to the latter than the former.

We do not dispute the laboratory findings: We think the data are robust (see Simion [2021] for defense of this claim against some recent findings). Rather, we suggest that these studies do not model real-world testimonial exchanges. Outside the lab, we are sensitive to agents' incentives to deceive us and to their reputations as informants (Mercier 2020). We punish informants who deceive us or whose confidence in their reports is misaligned with their evidence by excluding them from further cooperative interactions. Our inability to detect deception is unimportant when we can select informants who are unlikely to deliberately or inadvertently mislead us.

Our capacity to track reputation and to align incentives is a component of our broader systems of *epistemic vigilance* (Sperber et al. 2010), which are designed to protect us against misinformation. Epistemic vigilance mechanisms, at both the personal level and the sub-personal level,[7] are sensitive to genuinely good evidence of credibility. We prefer the testimony of those who share our values and of those who give evidence of competence. The testimony of the competent is more likely to be reliable than the testimony of the less competent (Mascaro and Sperber 2009; Sperber et al. 2010; P. L. Harris 2012). The testimony of those who share our values is less likely to aim at exploiting us (Levy 2019a). We thereby protect ourselves against fools and knaves.

The mechanisms of epistemic vigilance might be seen as constituting our epistemic immune system. When misinformation goes viral, we rely on these mechanisms to filter it out as a candidate for belief. Some testimony is *rationally* preempted for us (Begby 2021; Levy 2021): We place (almost) no credence in it, either because we regard the source as having no credibility or because it conflicts strongly with our settled beliefs. Some testimony is seen as requiring corroboration before it is accepted (e.g., testimony from outsiders; see Greco 2020); again, source or content (or, of course, both) determines the weight we place on it. Some testimony might actually be assigned a *negative* weight: There are circumstances in which testimony that *p* might increase someone's credence that not-*p*. (Suppose you're inspecting a house with a view to possible purchase. The realtor tells you that there's a lot of interest in the house and urges you to move quickly or risk missing out. Upon receipt of their testimony, your credence that there's a lot of serious interest in this property might reasonably fall: If there really was such interest, the realtor wouldn't be so keen on impressing the fact on you because they'd be confident of a sale in any case.)

It may be that our mechanisms of epistemic vigilance have yet to catch up with the contemporary high-choice, high-density, informational environment. These mechanisms evolved in small-scale societies where the opportunities for deception were fewer than today (Sterelny 2012). Perhaps cognitive mechanisms that were reliable in such environments now misfire because the world is far more epistemically hostile (Stanovich 2018). Most of us now live in an environment in which testimony often comes from informants with whom we will never interact and whose reputation is hard to monitor.[8] Nevertheless, that we possess such mechanisms seems beyond dispute, and the evidence that few people are taken in by bizarre misinformation suggests that they continue to protect us reasonably well against the formation of false beliefs.[9]

Even many cases in which people come to accept beliefs contrary to the scientific consensus arise because mechanisms of epistemic vigilance are functioning appropriately: In the right circumstances, such mechanisms filter out reliable testimony because the recipient has good reason to distrust it. As Adekola and Simion (2025) note, the datum that Scotland's Black, African, and Caribbean communities exhibited vaccine hesitancy during the COVID pandemic is better explained by these communities complicated history with Scotland's NHS than by irrationality. Scotland's Black, African, and Caribbean communities have been given reasons to be wary that Scotland's NHS was looking out for their community's best interests (Adekola et al. 2022). The lesson generalizes to the vaccine hesitancy of other communities: It's often most pronounced among communities that have been given little reason to trust health authorities (Goldenberg 2021).

1.2 The Bad News

So, there's lots of good news: People are not gullible and have robust defenses against bizarre misinformation. The good news ends here, however.

Although we're well protected against bizarre misinformation, we are nevertheless deeply epistemically vulnerable. Our evolved mechanisms of epistemic vigilance developed in light of this vulnerability: We need them because we've always been dependent on the division of epistemic labor and on tools and techniques that few (or no) individuals could develop for themselves from scratch (R. Boyd et al. 2011; Henrich 2015). Although these mechanisms filter out a great deal of bizarre misinformation[10] as well as testimony that comes from those who do not value our welfare, *by design*, they offer no protection against certain sorts of counterintuitive claims, opening us up to epistemic manipulation.

When testimony comes from those in our in-group, we are likely to accept it unless it is implausible by *our* lights. That leaves us vulnerable because our lights themselves are directly and indirectly shaped by our in-group.

What seems plausible or implausible to a person is shaped, in part, by their cultural background. In some cultures, the existence of ghosts and spirits is taken for granted; in others, relatively few people genuinely believe in them. Saying that my illness was caused by witchcraft strikes listeners as bizarre in some places and at some times but wholly unremarkable at others (Evans-Pritchard [1976] reports that the Azande believe that most illnesses are caused by witchcraft). There is also variation *within* cultures. Although members of each specific culture share a great deal in terms of central beliefs (more or less by definition), complex cultures always contain subcultures that may differ from each other and from the mainstream in their beliefs. Antivaxx beliefs, creationism, and climate change skepticism, for example, might seem much more plausible than scientific beliefs to many people, despite their living in a culture in which science is well established. When testimony comes from sources that our culture or subculture sees as epistemically authoritative, we tend to accept it—even when it is highly implausible by the lights of outsiders.

Indeed, the widespread acceptance of the science of vaccines, of evolution, and of climate change *itself* testifies to the power of testimony to overcome implausibility. None of these are intuitively obvious: far from it. Vaccination is highly counterintuitive (Miton and Mercier 2015), and even those with college-level education in evolution continue implicitly to reason about biology in terms that conflict with it (Shtulman 2006; Shtulman and Calabi 2013). That emissions can change the climate permanently is far from intuitive (and that the greenhouse effect can make *cold* weather more extreme seems bizarre). That people accept scientific testimony on all these topics demonstrates just how powerful testimony can be in bypassing mechanisms of epistemic vigilance.

The power of testimony to bypass these mechanisms is a feature of our cognitive systems, not a bug. Cultural evolution, through which our mechanisms of epistemic vigilance plausibly (albeit partially) developed, needs us to accept claims that we don't fully understand and which may strike us as bizarre. The solutions it hits upon to adaptive problems often include elements that seem superfluous but are, in fact, necessary. Consider the processes of detoxification of food that many cultures have developed—for example, the detoxification of cassava. Like many nutritious plants, cassava contains toxins as a defense against grazing animals. In humans, cassava consumption can lead to acute and chronic cyanide poisoning. Chronic poisoning is especially insidious: It develops slowly, causing developmental disorders and immune suppression, and its cause is far from obvious to those who consume cassava. Indigenous cultures, however, have developed preparation techniques to ensure poisoning is avoided.

For example, the Amazonian Tukanoans process cassava over a number of days. They first scrape the plant, then they grate it before washing it to separate the fiber and the starch from the liquid. These solids are then left to set for two days, after which it is safe to bake and eat them (Henrich 2015). These steps seem extraneous, and anyone attempting to reverse engineer cassava preparation might be tempted to skip them. In doing so, they run enormous risks.

Consider the fate of the 1860–1861 Burke and Wills expedition, which aimed to cross Australia from south to north and return. In April 1861, on the return journey, the already much depleted and desperate band encountered the Indigenous Yandruwandha people, who came to their aid with gifts of fish, beans, and damper (bush bread) made from the nardoo plant. In May, however, the Indigenous group moved off, leaving the party to fend for themselves. They spent the next two weeks camped next to a nardoo patch, preparing the damper for themselves.

They ground the sporocarps of the plant as they had the seen the Yandruwandha do, using some grinding stones they found. They made flour, mixed it with water, and baked it. According to Wills's journal, they consumed four to five pounds of the damper a day, which should have provided sufficient calories for them. It satisfied their hunger, but they became progressively weaker as the days passed. At least part of the reason for their decline is that nardoo is mildly toxic unless it is processed properly. The Yandruwandha people process nardoo to eliminate the toxins; the expedition did not see the need for the long processing (nor could they have replicated it if they'd tried). Burke and Wills were both dead a week after Wills wrote his journal entry on nardoo (Henrich 2015).

Cultural evolution works because it allows us to acquire cultural innovations even when, like Burke and Wills, we cannot see their point (there are many cases where Indigenous people give no explanation for their adaptive practices beyond "it is our custom" or even have mistaken beliefs about its necessity; see, e.g., Henrich and Henrich 2010). Comparative experimental work with human children and chimps demonstrates that humans are distinctive in copying models exactly, whereas chimps copy only those aspects of the technique that seems to be necessary for the goal (Nagell et al. 1993).[11] Adults are no more selective in what they copy when they regard the model as an expert (Lyons et al. 2007).

Our dependence on cultural evolution for flourishing entails epistemic vulnerability: When a source of testimony passes certain tests, we assign a high degree of credibility to it. We weigh their testimony by reference to markers of competence (where such markers adaptively vary from group to group), markers of benevolence toward us, and the plausibility of the testimony (Mascaro and Sperber 2009; P. L. Harris et al. 2018). How people balance these cues against one another may differ from case to case, but cultural evolution requires that when claims come from sources that appear competent and benevolent, we will accept apparently implausible claims (e.g., the claim that nardoo gruel must be consumed with mussel shells). It's a not a bug that renders us susceptible to believing apparently implausible claims, it's a feature.

How, in the face of this pervasive dependence on others, can we hope to achieve epistemic autonomy? Epistemologists often attempt to provide some sort of test or criteria to make up for the deficiencies of our mechanisms of epistemic vigilance. The literature on the so-called novice/expert problem is exemplary here (Goldman 2001; E. Anderson 2011; Guerrero 2017; Grundmann 2025). This literature asks which of the competing and conflicting experts we nonexperts should trust or, to put it in terms of mechanisms of epistemic vigilance, who we should trust when conflicting testimony comes from two or more individuals who possesses markers of competence. The resulting work is rich and suggestive, and we think the criteria it advances are genuinely helpful. However, they fall short—they *must* fall short—of providing us with grounds for choosing experts that bypass the need for antecedent trust. If you don't trust well to begin with, you're out of luck.

One of us made the case for the claim elsewhere (Levy 2022b; Levy and Varley 2024). Although we won't attempt to set it out in full again, we will briefly sketch some of the principal obstacles to assessing the reliability of competing (apparent) experts below. Before we do so, however, we want to point out that the difficulty is exactly what we should expect if we accept (as do almost all philosophers of science) that individual scientists are *themselves* ineliminably epistemically dependent on one another. If *scientists* can't dispense with trust, the same must hold (much more strongly) for laypeople.[12]

To say scientists are epistemically dependent on one another is to say that individual scientists regularly aren't in a position to check the data, tools, and methods produced by other scientists, even when they must rely on these data, tools, and methods. They may lack the capacity to check results for lack of time: Since scientific findings might take thousands of hours to produce across multiple laboratories, individual scientists cannot hope to check them (Hardwig [1985] is a classic reference for this claim). Often, though, they lack the skills to check the findings of other scientists or to understand how the tools they use were produced (Millgram 2015; Winsberg 2018; Gerken 2022): Science is highly specialized, and even in one lab, it's common for individuals to have skills that none of the other lab members possess (and can't possess, short of going back to school).

That's not to say, of course, that scientists can't detect fraud or incompetence. Statistical analysis can be analyzed for signs of what have come to be called "questionable research practices"—for example, suspicious clustering of results at the threshold of statistical significance (Ritchie 2020)—and scientists can analyze the logic of an experiment to discover whether these tools were applied correctly and whether the resulting data are interpretable or provide genuine support for the hypothesis. None of this requires that they bestow trust. But adjudicating between claims when there is no apparent fraud—for example, when scientists disagree about which sorts of evidence are most relevant to a question—takes genuine and field-specific expertise, to the extent it can be done at all. Experts who lack the *precise* set of skills and competences required for such assessment must go by the testimony of those they trust in order to choose. As Hardwig (1991) notes, scientists must trust other scientists, even in the absence of non-question-begging grounds for that trust.

Obviously enough, the novice is even more reliant on trust. Novices *can* sometimes uncover blatant fraudsters and cranks, and they can use the criteria put forward in the debate over the novice/expert problem to identify more and less reliable experts. But they can do these things only if they *already* trust well.[13] If experts can't identify non-question-begging grounds for trust, novices have no hope whatsoever.

There are two basic problems that together entail that novices cannot choose well if they don't trust well. First, although the criteria that philosophers such as Goldman (2001) and Anderson (2011) defend are appropriate and regularly helpful, they are also regularly mimicked (Guerrero 2017), and those who don't already trust well have no independent grounds for choosing the actually reliable sources over their mimics. Second, when an expert fails to score well in the light of one or more of the criteria, they can cite plausible excuses of just the sort that actually reliable experts themselves have recourse to.

Consider mimicry first, taking as our example Goldman's first criterion: dialectical superiority. Dialectical superiority is demonstrated (inter alia) by an asymmetry in the rebuttals that experts offer: If one expert more often rebuts the arguments of the other than vice versa, this tells in favor of the first expert, *even if the novice is unable to evaluate the cogency of the rebuttals* (Goldman 2001, 95). We agree that the inability of one side to offer a coherent-sounding rebuttal to the arguments of the other is suggestive. But it should be obvious that it doesn't help with sophisticated charlatans. They come well prepared. Sophisticated climate change deniers, those produced by right-wing American think tanks such as the Heartland Institute, have smooth and apparently coherent rebuttals at their fingertips. Indeed, the markers of dialectical superiority may *favor* them: They train for debating, not for science, and they are unconstrained by truth or even validity in their responses. Markers of dialectical superiority are quite easily mimicked. Other markers of scientific respectability require more effort, but motivated charlatans find it worthwhile. The merchants of doubt on climate change set up parallel institutes and journals and award parallel prizes (Oreskes and Conway 2011).

Of course, genuine scientists can adduce criteria on which their side does appreciably better than the charlatans. Anderson (2011, 146–147) points to citation counts and the impact factor of the journals in which scientists publish. This brings us to the second problem: The charlatans have a coherent excuse for their failure to score well on these criteria—one that doesn't look like special pleading. The journals won't accept their papers *because they're run by warmists*. They present themselves as brave outsiders, guided by the light of the truth and not acceptability to funding agencies and to mainstream scientists. They may cite the example of (apparent) iconoclasts against whom the scientific establishment closed ranks. Galileo's heliocentrism, Wegener's continental drift, and Warren and Marshall's theory of a bacterial cause for stomach ulcers were all vindicated, but only after decades of derision by the mainstream. Although we think that all three cases of iconoclasm are as much myth as reality (Levy and Varley 2024), it's undeniable that scientific assessment is influenced by values and not merely by the data.[14] If they don't trust well already, novices will be able to assess neither whether the climate change skeptics are the brave iconoclasts they present themselves as being nor whether the actual influence that scientists' values play in their work is appropriate.

1.3 Learning to Live with Epistemic Dependence

Although we do not take these reflections to be definitive, we take the forgoing to present a serious case for thinking that trust is ineliminable. That claim is the starting point of this book. In its light, we maintain that if epistemic autonomy is a realistic goal for agents like us, it must reckon with our epistemic dependence rather than attempt to eliminate it.

It's helpful here to compare debates over the novice/expert problem and related attempts to show that we can dispense with or non-question-beggingly ground trust with debates over global skepticism. Perhaps one day someone will produce an entirely satisfactory and non-question-begging rebuttal to the skeptic (although we doubt it).[15] In the meantime, most of us have concluded that we simply don't need Cartesian, infallibilist foundations for knowledge, not even for knowledge of contingent facts.[16] We've learned to live with our inability to rebut the skeptic who demands otherwise. Similarly, we believe, we need to learn to live with the impossibility of dispensing with or grounding trust.

Indeed, living with epistemic dependence is an urgent task in a way that confronting global skepticism never has been. We're very confident that the evil spirit imagined by Descartes is not actually attempting to deceive us. Global skepticism of that sort is a possibility we can't dismiss, but it's a bare possibility—hence the lack of urgency to address it. Simulation arguments and Boltzmann brain arguments (cf. Dogramaci and Schoenfield 2025) purport to show that other kinds of skeptical scenarios are likely (Bostrom 2003), but they depend on premises that are eminently disputable (K. R. Harris 2024). They certainly do not succeed in showing that skeptical hypotheses are actual. Harmful epistemic dependence is no bare possibility, however: It is an all too real actuality. We live in a world in which there are multiple malicious or merely self-interested agents who attempt to influence our beliefs for their benefit and not ours. It's relatively easy to live with the failure to rebut global skepticism. Learning to live with epistemic dependence is *hard*.

It is the harder because there is no one-size-fits-all response to the problem of epistemic dependence. Instead, we need strategies that are context specific. We may need different strategies for living with our dependence on experts to those we need to deploy to live with our dependence on technological scaffolding, for example. We must investigate different contexts in detail to assess the degree to which we need different strategies in each. We must also come to understand how we can maintain the bases for our self-respect as decision-makers—how we can be epistemically autonomous—in a world in which dependence is pervasive and ineliminable.

Given the number of different contexts in which the problem recurs, and the different forms it can take, we do not aim to offer either a comprehensive theory of epistemic autonomy or an overview of the many strategies that could be deployed in its service. We hope to make progress on both fronts, but the progress will be partial. We will consider some of the more important contexts in which we must learn to protect autonomy while recognizing our dependence, and come nearer to understanding what an adequate account of epistemic autonomy—as we see it, *social* epistemic autonomy—will look like. We do not dream of a final theory. We offer the book as a contribution to a debate that welcomes many voices.

2 Hijacked Attention

Until recently, attention has not had the attention it deserves in philosophy. Philosophy of mind was first to begin making up for this neglect (Mole 2011; Watzl 2011; Wu 2014; Dicey Jennings 2020). More recently, the importance of attention (and salience, to which it is closely related) for normative theory has increasingly been recognized. One strand of research extends the Lewisian "scorekeeping" approach (Lewis 1979) to understand how hate speech makes prejudiced assumptions salient in a conversational context (McGowan 2003; Simpson 2013; Lepoutre 2021; McGowan 2022; for criticism, see Fraser 2023). A different (although compatible) strand grounds prejudice (Munton 2023), virtues and vices (Gardiner 2022; Watzl 2022), and certain sorts of harms (E. Whiteley 2022) in patterns of attention, whereas yet a third strand focuses on how emotions capture our attention, prompting the search to seek out facts or considerations that might justify the emotion (Brady 2016). Epistemology's traditional focus has been on knowledge and belief. If Sebastian Watzl is right that "attention and salience are explanatorily central in an account of the mind and in the explanation of experience and behavior to a degree that is, as least, roughly commensurate to the explanatory role of belief" (Watzl 2022, 92), perhaps epistemology should follow other areas in making attention a focus.[1]

In this chapter, we will focus on one particular harm that falls within the broad purview of epistemology and that can only be understood by focusing on attention.[2] We will argue that the harms of *implausible* misinformation arise very significantly from its effects on attention. We will also argue that this is bad news: We are less well equipped to protect ourselves against inappropriate attentional capture than against attempts to induce false beliefs in us. We possess a suite of capacities for epistemic vigilance, but we lack equivalent capacities to protect us against pernicious influences on our attention.

2.1 Misinformation, Plausible and Implausible

That misinformation may be harmful is obvious. We've already set out the central reason why: We humans are epistemically dependent animals. We rely on the testimony of others pervasively: We would know very little about the world, even the intimate world of our own family, without testimony from others. Moreover, this dependence is central to our flourishing. Because humans live in such a diversity of environments and face a diversity of challenges, we're dependent on testimony about the local way of doing things to flourish. Ways of behaving that are adaptive on the African savannah are very different from those that are suited for Lagos or Karachi or Copenhagen; we must acquire the local norms and the local practices if we are to do well.

As we saw in the first chapter, successful local practices may—moreover—be complex and counterintuitive. We cannot expect to understand just *why* it's important to do things *this* way and not that. Under the right conditions, we are therefore disposed to both practical and propositional beliefs that seem very strange to outsiders (Levy and Alfano 2019). Our flourishing depends on our acquiring beliefs that are highly counterintuitive. It's hardly surprising, then, that we may come to acquire some very odd beliefs as a consequence of misinformation.

As we also saw, however, there is strong evidence that when people report bizarre beliefs, they are often either insincere or mistaken. Psychologists and the popular press sometimes look for explanations of these sorts of reports in our apparent gullibility (D. T. Gilbert et al. 1993; Robson 2016). In the light of evidence that we're designed to accept information from others, including highly counterintuitive information, the fact that we *rarely* believe the more bizarre claims conveyed to us via testimony is a more pressing target for explanation. Why *aren't* we gullible? Why do relatively few people genuinely believe the weirder conspiracy theories?

In the light of the evidence from cultural evolution briefly reviewed in the first chapter, we think it's likely that it's not the implausibility of the testimony alone that explains the fact that few people believe the fantasies of QAnon or that the planet Earth is flat. Were agents to receive testimony from those they regarded as genuine epistemic authorities with that content, we think they *would* go on to form the correlative beliefs. That they don't is, we think, good news: It shows that even those people who see themselves as (in some sense) on the same side as the misinformation superspreaders (Baribi-Bartov et al. 2024) don't see them as epistemic authorities. They may see them as passing other tests for reliability (e.g., sharing their values), but that's not sufficient to accept such claims.[3] We think it's likely that the very fact that people retail bizarre conspiracy theories disqualifies them as credible sources of information, even by the lights of those people who seek them out and delight in them.

On the picture we have in mind, the implausibility of a claim, its departure from the common run of our experience and from the consensus view of the world, raises the bar for its acceptance. This is just what we should expect: As Bayesian approaches to belief update make clear, whether we should believe that a claim is true, given some body of evidence, depends, in part, on its prior probability. The testimony of a somewhat reliable source is sufficient for believing that some expectable event occurred, whereas testimony from the same source may be insufficient for believing that some very unlikely event occurred.

When testimony concerning some event or claim that is expectable comes from a source that is not obviously biased or incompetent, we tend to accept it. We should expect, then, that misinformation quite routinely causes false beliefs. But when stakes rise, agents are more careful. They hesitate to form full beliefs. The major doxastic harm of misinformation, we suggest, arises in situations where agents act on credences that fall short of belief because they are under time pressure and unable to verify the testimony quickly. For example, fake news can cause significant changes in stock prices (Gold and Stelter 2025; Arcuri et al. 2023). Given the pressures traders work under, they cannot avoid responding to half-plausible rumors. Scams, too, may induce behavior without inducing belief (Levy 2025). In these sorts of cases, agents are brought to act in something like the way they would act were they to believe the misinformation without them actually believing it.

For most of us most of the time, however, implausible misinformation causes little direct harm. It neither changes beliefs nor induces behavior of a sort that directly reflects the content of the misinformation.[4] When it matters to us, we do not accept it, or we accept it only because it makes little difference to us and how we act. Breathless headlines notwithstanding, we're too discerning often to be taken in by unbelievable misinformation.

2.2 Misinformation and Attention

Should we conclude, then, that no matter how harmful plausible misinformation may sometimes be, implausible misinformation is harmless? We suspect not: Although it is rarely believed, implausible misinformation likely gives rise to a number of harms. Chief among these, we will argue, is how such misinformation occupies and skews our attention.

Roughly, the capacity to attend is the capacity to make some representation or content the focus of concerted cognitive resources; attention is the state in which resources are brought to bear in this kind of way. Attention is neither necessary nor sufficient for belief: We believe all kinds of propositions to which we are not (currently) attending, and may never have attended, and we can attend to a proposition without believing it. (We regularly attend to fictions without being inclined to believe them.) Attention and belief are not merely different, in the sort of way in which, say, memory and affect are different. They are also, plausibly, deeply different, belonging to different orders of psychological states. Although both may be governed by norms, belief is also partially norm *constituted.* If a state differs too significantly from belief paradigms, it is not a belief at all, any more than a game in which players win by throwing pieces at one another is chess. Attention comes in degrees all the way along, as it were: There's no such thing as attending too little to count as attending at all. Attention is also plainly under direct voluntary control in many instances (although far from all). It is, at minimum, an open question whether belief is ever under direct voluntary control. (Cards on the table: We reject doxastic voluntarism.)

Since attention and belief are deeply different, it is unsurprising that they often dissociate. We see this dissociation with regard to misinformation as much as anything. We attend to misinformation much more than we believe it.

Implausible misinformation is especially likely to be attention-grabbing. What makes a report attention-grabbing is not well understood, but there's no doubt that being surprising is centrally important. On the influential predictive coding account of the mind, attention is a mechanism for making surprising input more precise in order to be able to assess it better

(Feldman and Friston 2010; Hohwy 2013). The properties that render misinformation implausible strongly are conducive to making it surprising: It's the very fact that it is out of the run of common or expected occurrences that makes it implausible. Implausible misinformation is often attention-grabbing for other reasons. In particular, it often arouses strong emotion and triggers coalitional feelings (Stubbersfield 2023).

Of course, true reports and plausible misinformation can also be attention-grabbing. They, too, may trigger coalitional feelings and arouse strong emotions. Reports of behavior that arouses indignation are attention-grabbing, whether the behavior is plausible or implausible, true or false. But bizarre misinformation is especially likely to have these properties. It is *produced*, after all, to be memorable; its production is shaped not by the demands of truth (which is often boring) but rather the demands of storytelling. Bizarre misinformation that is widely repeated is most especially likely to be attention-grabbing for obvious reasons: its success is more or less decisive evidence of its attention-grabbing nature (Acerbi 2019).

Now, there is nothing wrong with being attention-grabbing per se. Some events and some reports *should* be attended to. Attention allows us to plan for the future and cope with the unexpected. When something surprising really happens, we attend because its occurrence suggests we need to update our model of the world. When something that arouses our indignation occurs, we attend because justice matters to us. When something happens that arouses our coalitional feelings, we attend because we or those we care about may be threatened. But attention may be shaped perniciously too. We may be brought to attend to events (real or otherwise) that do not deserve our attention (because they are relatively trivial, for example, or because they are unrepresentative). Perhaps we could hope to formulate a general account of what makes an event or a report worth attending to; perhaps, on the other hand, such an account would be irreducibly normative, such that it would be of little more help than "attend to all and only those events that are worth attending to." Fortunately, we don't need such an account here: We can work with clear examples in which attention is captured by events or reports that we would be better off not attending to or which should go unattended. Not only is bizarre misinformation especially likely to be attention-grabbing; it is also especially likely to attract *unwarranted* attention, since its claims are neither true nor up for serious consideration as candidates for truth.

Let's turn now to the harms of this sort of attentional capture. We've already mentioned some of the ways in which attention is normatively central: to the development of virtues and vices, to the constitution of prejudices, to certain sorts of epistemic injustices, and so on. We will introduce the kind of harm we're concerned with here by comparison to another we've also already mentioned: how making something salient[5] can introduce pernicious presuppositions into conversations (McGowan 2022, 2003; Simpson 2013).[6]

In groundbreaking work, Mary McGowan (2003, 2004) has argued that speech can be oppressive when it introduces presuppositions into conversations that—*because* they are not asserted—are difficult to challenge but nevertheless alter what is conversationally acceptable. Such oppressive speech may work by making harmful stereotypes salient. Even making race or gender inappropriately salient may change the conversational score and constitute a kind of sneaky norm enactment (McGowan 2022)—for example, when a physician's race is made salient, her credibility might be undermined.

It's important to emphasize that this kind of harm is independent of belief. In McGowan's example of the well-meaning man who makes his colleague's race salient, no one need believe that the colleague is any less of a medical expert after her race is made salient. His assertion nevertheless alters what is conversationally appropriate. Perhaps broadly distributed misinformation may have similar effects on the public (and private) conversations that constitute political exchange; we leave it to those with expertise in speech act theory to assess whether the framework can be scaled up in this manner. We want to point to a related but distinct phenomenon: how making a topic salient changes the *shape* of a debate. It changes what is up for discussion and how it should be discussed.

The same properties that make implausible misinformation memorable make it salient and thereby change the shape of the debate. Once again, these effects may occur independently of whether it is seen as credible.[7] For example, when misinformation is introduced, it may become incumbent on some debate participants to address it, even though they don't take it seriously—and even if *no one* really takes its seriously.[8] How many people genuinely believed Donald Trump's claim that Haitian immigrants were kidnapping and eating dogs and geese (Liddell 2024)? Although some people probably did go on to accord it a significant degree of credibility, we suspect most did not (perhaps including Trump himself). JD Vance, who seems to have been instrumental in bringing the claim to Trump's attention, distanced himself from it at the same time as he introduced it (Chait 2024). A lack of belief did not prevent it becoming a central topic of discussion, including—or especially—among those who explicitly dismissed it as nonsense. As this example illustrates, when such misinformation comes to be widely attended to, it tends to reorient the conversation away from productive terrain (What facts and what measures can we agree on?) and down unproductive alleyways.

Making misinformation salient may therefore be an effective tactic for disrupting debates. This makes misinformation a handy tool for those who don't want their side to reach any sort of compromise with the other side. For example, misinformation portraying Democrats as Satanists (or Republicans as flat-earthers) might be introduced to ensure discussion breaks down. If the accusation could simply be ignored, since the claims are so outlandish, it might do no real harm, but its attention-grabbing nature makes it hard to simply dismiss (as the success of QAnon, to which the Satanism claim is central, at capturing the attention of many newspaper columnists and pollsters illustrates). If we are to continue to engage with those who pretend to take such claims seriously, or even those who are allied with those pretenders (including Trump himself; Ibrahim 2020), we might be forced to confront the claims in one way or another. Indeed, many outlets have devoted articles and think pieces to these claims (e.g., Donegan 2020; Caldwell 2021; Venkataramakrishnan and Murphy 2021). Such claims continue to obtrude into consciousness and make productive conversation difficult.

Finally (but very importantly), attention is a limited resource. We can only attend to a certain number of things at once. Attending therefore has opportunity costs. The conflict in Ukraine slipped off our radar while the war in Gaza raged; at the time of writing, Trump has riveted the world once more, and neither conflict is paid much attention. What we attend to determines what our priorities are. The fact that attention is a limited resource ensures that those who can capture our attention can exert a powerful influence over our priorities. That's a truly awesome power.

It's an open question, we think, whether these effects are well characterized as changing the conversational score. At least, their effects do not turn on the introduction of any presuppositions. They change the shape of public debate independent of any such effect.

In the next section, we will argue that the effects misinformation has on our attention are especially difficult to address. For good reasons, it is very difficult to avoid attending to what is made salient; attention is far more vulnerable than belief.

2.3 The Attention/Belief Asymmetry

It may be, as some researchers have suggested, that anxiety about misinformation arises in the main from the concern that others are gullible (Altay and Acerbi 2024), and if it were more widely known that bizarre misinformation is rarely believed, the anxiety would diminish. Misinformation might therefore become less salient and recede, to some degree, from public view. In this section, however, we will argue that token instances of bizarre misinformation are very effective at capturing our attention, no matter what we believe, and that we lack good defenses against inappropriate attentional capture.

Our vulnerability to attentional capture contrasts with our relative security against false beliefs. As we emphasized in the first chapter, our deep and pervasive reliance on testimony, explicit and implicit, requires that we have an epistemic immune system: a set of mechanisms for filtering information for credibility. Epistemic vigilance mechanisms are far from foolproof, but they do an excellent job of protecting us against implausible misinformation. Its very implausibility ensures a high degree of scrutiny when it matters to our lives. Few people are genuinely taken in by bizarre misinformation. But we are much more vulnerable to pernicious attentional capture.

Our epistemic immune system consists in a suite of mechanisms that are (plausibly) adaptations to protect us against false beliefs. Our attentional system is designed to opposite ends: It consists in mechanisms that are designed automatically to bring us to attend to information that is important to our goals. This renders us vulnerable to attentional capture. Attentional capture occurs, centrally, because others attending to something provides us with a genuinely good reason to attend to it, either because it's important for us (for a variety of reasons) to know what others attend to or because their attending to a topic confers significance on it, even when it would not deserve attention were it not attended to. We will now expand on how this occurs.

2.3.1 Cognitive Architecture

Part of the reason we lack attentional vigilance mechanisms is that our deployment of epistemic vigilance *requires* that we attend to what our conversational partners attend to. We engage in mind reading to deploy epistemic vigilance: Assessing others' reliability as testifiers and their motivations to aid or to deceive us requires that we attempt to detect their beliefs, desires, and goals (Mascaro and Sperber 2009). But mind reading requires attending to what other agents are attending to.

Consider the false-belief task, the single most studied test of mind-reading capacity. In the false-belief task, participants watch one agent move a desired object from where it was originally placed to an unlikely location (e.g., children might see the agent remove the chocolate from the candy box and place it in a pencil box). They are then asked to predict where an agent who saw the object being placed at the first location, but not its subsequent relocation, would look for the chocolate. Children begin reliably to pass this test around the age of four; prior to that age, they predict that the agent will look where the chocolate actually is (Mitchell and Lacohée 1991; Wellman et al. 2001). Although it is now controversial whether the test reliably tracks developmental trajectory (Heyes 2014), it's the logic of the task, not its reliability, that matters to us here. What the task nicely brings out is how the attribution of false beliefs to an agent requires tracking what they attend to: The agent falsely believes the object is in place *a* because she did not see that the object was moved to *b*. Accordingly, mind reading depends on attending to what others are attending to, including what we attend to jointly, using cues such as eye gaze (Baron-Cohen 1997). Being epistemically vigilant *requires* that we attend to what others attend to.

2.3.2 The Significance of the Attended

Animals pay attention to what other animals, including those outside their own species, attend to because they have a great many threats and opportunities in common. For instance, members of one species may be attentive to the alarm calls of other species because they have predators in common (Magrath et al. 2015). What is true across species holds more strongly *within* species, since the overlap in threats and opportunities is much greater within than across species. So, we can be expected to orient toward stimuli that others attend to. Members of the out-group may be mistaken about many things, but insofar as we share an environment (where an environment might be physical, cultural, social, or virtual), *their* attending to something is a reason for *me* to attend to it.

2.3.3 Attention Confers Significance

Perhaps counterintuitively, the fact that something is widely attended to just *is* a reason to attend to it, even if it wouldn't be worthy of attention otherwise. *That something is the focus of attention* makes it important (Gardiner 2022). There are at least two mechanisms that make the attended ipso facto important. First, we're cultural animals; culture isn't optional for us but genuinely formative. When something is very widely attended to in our culture, it becomes a reference point for us—something in relation to which we're required to orient ourselves. Although there are subcultural groupings that are genuinely oblivious to what the mainstream attends to, this obliviousness is both a symptom and a cause of alienation: Such groupings are marginalized. They are unable to participate fully in the conversations whereby we make and remake ourselves. When obliviousness is broad enough, such subcultures suffer not so much from testimonial injustice—where their voices go unheeded or are discounted—but rather from a lack of a voice at all (in fact, this sort of genuine obliviousness is usually a symptom of a deeper alienation and deprivation).[9]

A second mechanism whereby attention produces something as worth attending to might be illuminated by comparison to what Quine and Ullian (1978) call the "web of belief." Some beliefs are more central to the web and others more peripheral. Those on the periphery have fewer linkages to other beliefs, and may alter or atrophy without ramifications for the web as a whole. Those that are more central are more densely connected, and alterations to them have broader effects. Attending to a claim tends to make it more central to the cognitive web (which extends beyond our beliefs) and tends to bring it about that it is more densely connected to other attitudes and views—including many with no obvious logical connection to it. In virtue of having been the object of sustained attention, almost any arbitrary content can come to have real significance for us.[10]

2.3.4 The Dynamics of Debate

Insofar as we engage with the out-group at all, it is difficult to avoid a focus on what they focus on. This is both a condition of engagement and a dialectical necessity. It's a condition of engagement: We can't be said to be engaging with the other side at all if they keep insisting on some point we refuse to address. It's a dialectical necessity: Bystanders might reasonably conclude that the issue is an important one and that our silence on it indicates that we have no good response to it.[11]

Consider an example from outside the domain of misinformation (strictly speaking). It has become a sport of right-wing media, at least in the UK, to ask prominent politicians to define "woman." The aim, of course, is to put politicians on the left in an impossible position: Either they refuse to answer the question, or they give an answer that will be fodder for headlines ("Keir Starmer Unable to Define a Woman"; Bet 2024).[12] Whether it's important to be able to answer this question is debatable. But the very fact that the right in the UK has made it central ensures that it's hard to set it aside: Conversations on feminism or trans inclusion are easily derailed by the insistence we settle whether trans women are women. A refusal to debate the question might be seen by bystanders (and those who insist on the debate) as an admission that we have no good response (as indeed the press gleefully claims; Penna 2022). Insofar as the other side insists on the question, failing to respond ensures the breakdown of conversation.

Of course, we might be quite happy to break off the conversation with those who cynically weaponize this sort of question. We might think they're unlikely to be productive conversational partners in any case. But the success of a group at making a question salient infects conversations that do not involve them. The issue is obtrusive whenever issues of the obstacles confronting women or of trans inclusion are raised. Bystanders and (naive or ill-motivated) interlocutors can exercise a kind of heckler's veto: The salience of the question means that its easily raised, and that once raised, it tends to crowd out more productive topics.[13]

We've identified four different kinds of reasons why we're vulnerable to attentional hijacking. When individuals with a prominent platform or significant sections of the population publicly attend to a person, an event, or a claim, potential conversational partners are also forced to attend to it. We must attend in order to engage in mind reading (and thereby filter testimony reliably), because attention is evidence that what is attended to is important, because attention *confers* significance, and because we cannot engage in debate without attending to what our interlocutors insist on attending to. Attention is mandatory, or near enough. In this respect, it contrasts strongly with belief. Our epistemic immune system protects us from believing what others believe when we assess it as unlikely to be true; *that they believe that p* does not bring *us* to believe that *p* when *they* are unreliable, by our lights. But *that they attend to x* does bring us to attend to *x*.[14]

We are therefore uniquely vulnerable to attentional capture. It is a powerful force: Donald Trump owes his presidency to his capacity to attract attention, both from those who support him (in large part *because* he's outrageous) and from those who oppose him (as well as the dwindling number of those who neither support nor oppose him). His ability to attract attention is reflected, for example, in the enormous amount of media coverage he receives. In the 2016 election, for instance, his primary campaign received more coverage than those of all the Democrat candidates combined (Stelter and Olshansky 2015) and more than twice the coverage of Hillary Clinton once she was the candidate (Confessore and Yourish 2016). *We* don't want to be talking about Donald Trump, but his capacity to dominate attention makes him inevitable.[15] We lack effective defenses against this sort of attentional capture. But it's very harmful: It might even threaten democracy.

2.4 Attentional Capture and Autonomy

There are many kinds of harm, and many of them have little to do with autonomy (except insofar as they undermine the kinds of conditions in which autonomy is best cultivated and maintained). Is the attentional capture characteristic of implausible misinformation a threat to epistemic autonomy?

Attentional capture is interesting, we maintain, for many reasons, but one reason is that it fits uneasily into existing understanding of epistemic autonomy. Although some thinkers recognize that being epistemically autonomous requires the self-governance of our "intellectual lives" generally (Matheson 2024, 267), its primary locus is held to be doxastic: Being epistemic autonomous requires that we make up our minds what to believe. As Goldberg puts it, "An epistemically autonomous subject is one who judges and decides for herself, where her judgments and decisions are reached on the basis of reasons which she has in her possession" (2013, 169). Attentional capture is interesting because it constitutes a non-doxastic threat to epistemic autonomy.

As we've seen, that others attend to something is a reason for us to attend to that same thing. But we do not attend to what they attend to in light of that reason; our attention does not reflect our assessment of what we have most reason to do. Our attention is deployed more reflexively than that. Indeed, we may judge that *that issue* is not worth our attention; we're forced to attend in any case. The culture wars turns, in large part, on questions that do not deserve much attention in virtue of their role in human life or their intrinsic significance. It's really not very important, in the grand scheme of things, whether the 2016 female-led *Ghostbusters* reboot or the 2025 *Snow White* remake are good films. Both became culture wars cannon fodder, however, and people who had better things to do with their time found themselves needing to have opinions on them.

Insofar as attentional capture leads agents to pay a great deal of attention to matters they judge to deserve little attention, it appears to threaten our autonomy over our intellectual lives. There are, however, grounds for thinking that the offense over our autonomy is relatively small compared with offenses against autonomy that are doxastic. It is the proper function of attentional mechanisms to orient toward whatever our attention is called to. "Look," someone says, "cows!" And you look, even if you take cows to be not worth attending to. It's hard to see any offense to autonomy in this sort of case. At any rate, any offense pales in comparison to the truly egregious infringement on epistemic autonomy that would be brought about by someone bringing you to believe something you took yourself to have no good reason to believe. Attentional capture can be grounds for serious complaint in virtue of its broader effects on our cognitive, and moral and political, lives, but it is far from clear it involves an infringement on our epistemic autonomy.

One salient difference between attentional capture and the inculcation of beliefs (without the presentation of evidence for those beliefs) is their relative durability. Although beliefs can be transient (consider perceptual beliefs about the mundane world around you, automatically formed and almost immediately forgotten), they tend to persist. They move into the background and persist as dispositions that shape further thought. Attending is typically transient. You glance at the cows, but having judged them of no interest or importance, you look away. You may be irritated by the interruption, but if it's a one-off, you won't make a big deal about it.

It is its relative durability, we think, that explains why attentional capture need not be an offense against epistemic autonomy to anything like the same extent as the inculcation of belief without the presentation of evidence. The fleeting induction of a belief would also fail to be a serious offense against intellectual autonomy. Conversely, when attentional capture is durable, it too can be a serious offense against autonomy.

Just as the inculcation of persisting belief is an offense against intellectual autonomy only when the belief is not caused by genuinely sufficient evidence, so too persisting attentional capture rises to the level of a significant offense against epistemic autonomy only when we judge the attended content to be unworthy of attention. When Bilbo draws Thorin's attention to the fire-breathing dragon, Thorin's attention is lastingly captured for the duration of the dragon's presence and beyond (assuming he escapes its clutches). Just as beliefs recede from consciousness but persist as dispositions, so Thorin may be disposed from that moment on to be vigilant for signs the dragon is nearby. Thorin has no complaint against Bilbo in this sort of case—quite the opposite.

Attentional capture is lasting when cues to attend succeed in changing the social landscape. Enough people must be disposed to bring it up often enough that it becomes salient to all of us, including those who judge it unworthy of attention. You might not think it's worth attending to this or that culture wars kerfuffle, but the knowledge that you will be called out for your privilege, or branded a Marxist, for expressing enthusiasm about some movie (for instance) is salient to you. It might act to dampen your enthusiasm, or at least your wish to express it. In these sorts of circumstances, you may have a legitimate complaint that your intellectual life has been shaped by others against your wishes and without your consent.

For all that, we don't think the most serious harms that arise from attentional capture are best understood as turning on how it infringes epistemic autonomy. Significant effects on epistemic autonomy depend on it having lasting effects, and these lasting effects require that it changes the social landscape. It is these changes to the social landscape that are most significant, morally and politically, rather than the way in which it constitutes an infringement of our epistemic autonomy.

2.5 Conclusion and the Way Forward

Bizarre misinformation is believed much less often than it is asserted or shared. Because it's not often believed, it doesn't often give rise to the harm most often feared from it. Those who share it and endorse it haven't really checked out from our common reality; they aren't likely to engage in the harmful actions that would be licensed by beliefs like *those*. But it's far from harmless.

Misinformation has properties that give it an edge in cultural epidemiology: Successful misinformation goes viral because it's memorable and attention-grabbing. It's attention-grabbing for all of us—those who find it congenial in some manner (even if not plausible) and those who find it repulsive alike. By capturing our attention, it changes the shape of debates and debate dynamics, and—perhaps most importantly—shapes our priorities. The power to capture our attention is an awesome one.

We've suggested that there's a deep asymmetry between belief and attention when it comes to our capacity to filter out what we assess as unjustified or otherwise bad. If we're right, it would be foolish to rely on our cognitive capacities to counter the capture of attention or the way in which it changes the shape of discussion. Our best bet, we suspect, is to look beyond such capacities, to social and institutional responses to unwarranted attentional capture.

Of course, individuals are not entirely powerless to counter attentional capture. Since attention is a limited resource, we can crowd out unwanted foci of attention simply by attending to the issues or questions we take to be really worth attending to. But the limitations of this approach are obvious: many of the harms of attentional capture stem from the way it shapes debate and public inquiry, and the lone individual's resistance doesn't begin to address these harms. We need a critical mass of people attending together to force the topics that really matter to continue to attract attention. That requires joint action.

It also requires that we attend to institutional and social mechanisms that govern who has more or less power to demand attention (to themselves or to what they attend to). The landscape here is complex, with subtle interactions between content and identity. For example, those with historical advantages, stemming from race, gender, and class, have greater attentional power in general, but other groups may have an attentional power with regard to certain topics (Black linguistic innovations are especially likely to shape the future development of slang in the United States, for example; M. Adams 2012). We urgently need mechanisms that ensure fairer allocations of attention and attentional power. When individual capacities fall short, we must compensate with social mechanisms that ensure that such capacities are not required.

We might, for example, develop better mechanisms for measuring how representative an opinion is. Consider "trending topics" on X (formerly known as Twitter). That a topic is trending is a reason for other users to attend to it. But the fact that a topic is trending might represent far fewer people attending to it in a first-order fashion than we tend to think. Further, and probably more significantly, much of the engagement is likely bot driven. Bots may be programmed to amplify certain topics, or they may simply respond to what others attend to, causing a snowball effect. Better measures of how much attention is genuinely first order—reflecting people's assessment that the topic is worth attending to—and how much is bot driven might reduce the power of attentional hijacking. Social media is often a hijacking amplifier: We need to develop algorithms that dampen such hijacking.

As we've also seen, though, certain topics have an edge when it comes to capturing attention, independent of who possesses attentional power. That fact ensures that bad actors have an advantage: Outrageous claims, especially those that are morally and politically charged, easily capture attention. We might want to focus attention on the issues that matter, but agents who want our attention elsewhere—anywhere but *here*—can capture the spotlight with relative ease. We face an uneven battle in developing social and institutional mechanisms for fighting back.

3 Group Doxastic Autonomy

In ordinary conversation and in public forums, people often attribute beliefs to groups. The Republican Party believes that Joe Biden was bad for America. Oxfam believes that widening inequality hurts the worst off. For a long time, mainstream philosophical opinion regarded such talk as metaphorical (see Quinton [1976] for a representative view), but today, philosophy is much more sympathetic to the idea. Although we argue that the conditions under which beliefs may veridically be attributed to groups are more demanding than others have thought, we are on board with the emerging consensus. On the novel account we develop here, groups that are integrated in the right sorts of ways and whose output is responsive to evidence can count as genuinely believing these outputs.

We argue that group members need not (and often will not) share the beliefs of their group. Our account is, thus, *non-summativist*—that is, it allows that groups may have beliefs that are not a function of the beliefs of the individuals that constitute it. We argue that our account is preferable to other non-summativist accounts. We do not altogether rule out a disjunctive account, whereby group beliefs might be realized summatively *or* non-summatively, but we give reasons for thinking that our non-summative account is preferable to the competitors.

Our primary aim in this chapter is not merely to defend this account, however, but also to explore its implications for epistemic autonomy. The autonomy of groups is a major concern of political and legal theory. The right to self-determination of peoples is enshrined in international law; self-determination is near synonymous with autonomy. But groups can be self-determining only if their beliefs are autonomous: They lack the capacity to determine their own development if their beliefs are imposed on them by others. Groups may lack certain protections against paternalistic interference if they are not autonomous in their beliefs.

We approach the question of the autonomy of group beliefs by asking about the relationship between the autonomy of individual members' beliefs and the autonomy of the group's beliefs. We ask, that is, whether the autonomy of group beliefs depend, in some way, upon the autonomy of the beliefs of group members (or, perhaps, on their autonomy as agents). Do heteronomous group beliefs supervene on individual members' heteronomous beliefs? Or can the two dissociate? We show that on all theories (not just our own), group belief may be autonomous, even while the relevant beliefs of group members (i.e., their beliefs about the very same topic) are heteronomous, and that on some theories, group heteronomy may coexist with individual-level autonomy. Although we will defend a (non-summativist) account of group belief, we also aim to map out the options. We will treat our own account as one among others in the latter part of the chapter, exploring its implications alongside the implications of summativist and joint acceptance accounts of group belief.

3.1 When Do Groups Believe?

Although it is no longer fashionable to see attributions of beliefs to groups as merely metaphorical, it's very plausible that such talk is often very loose. We're concerned with group belief, with "belief" understood in a strict sense. How do we sort the metaphorical sheep from the literal goats?

Genuine group beliefs will be states that play the same functional role as the genuine beliefs of individual agents. In this section, we will set down plausible functional conditions for genuine individual belief. The criteria we adopt allow us to home in more precisely on group beliefs and to distinguish group beliefs from other states that might aptly be attributed to groups.

Neil Van Leeuwen (2023) has identified four criteria we see as plausible conditions for counting as a genuine belief: involuntariness, no compartmentalization, cognitive governance and evidential vulnerability.

- Involuntariness, beliefs are forced on us by the world; we don't get a choice whether to believe a proposition.[1]
- No compartmentalization, beliefs govern behavior in all contexts in which they're relevant, at least insofar as they're salient to the agent.

- Cognitive governance, beliefs govern inference and not merely behavior. This governance even extends to beliefs about counterfactuals and imaginings.
- Evidential vulnerability, the involuntary adoption and update characteristic of beliefs occurs in response to evidence bearing on their truth and (more or less) nothing else.[2]

Van Leeuwen develops his criteria with the aim of building a case that religious attitudes are typically not beliefs.[3] Religious attitudes fail all these tests, he maintains. They are not involuntary; rather, they are acquired for their pragmatic benefits. They are compartmentalized to certain settings and do not affect behavior outside those settings. They do not affect factual inference, and they do not respond to evidence. Van Leeuwen argues that it is only if we deny that religious attitudes have these properties that we can make sense of the behavior of those who claim to be believers. For example, the member of an evangelical church who interprets the shock his toaster gave him as evidence of demonic activity will be careful to have it repaired—and not (just) exorcized—prior to using it again.

We suggest adding a fifth criterion to Van Leeuwen's four. In addition to involuntariness, no compartmentalization, cognitive governance, and evidential vulnerability, genuine beliefs are *poised for action and inference* (call this the "poise condition"). This criterion is required to make a principled distinction between our dispositional beliefs and those beliefs we would come to hold were we to think carefully enough. Our dispositional beliefs are those we would assent to immediately were we asked, even when they've never been occurrent (such as "bananas are not good building material"). We assent to them immediately because they're entailed by beliefs that are easily accessible. Those (counterfactual) beliefs that are not poised for action are also entailed by our beliefs, but the inferential relations are hard to see: At best, it would take time and effort for us to recognize we're committed to them.

States that are poised for action and inference tend actually to guide us when their content is relevant. We don't need to be prompted to think about them. We don't try to use bananas to build a house because we dispositionally believe that they're not good for that purpose. In contrast, states that we would have were we to think about the question they bear on long enough and carefully enough do not guide our actual behavior and inference. In detective stories, the clues to who the murderer is may all be presented to the reader, but the denouement comes as a surprise nevertheless. Sometimes, we fail to "put two and two together," but we do dispositionally believe that two plus two equals four. The kind of case we're thinking of is more like 33,873×0.81. Although a unique correct answer to that question is entailed by facts we know, the answer is inaccessible to most of us without effort and does not guide behavior and inference under normal circumstances. We dispositionally believe that bananas are not good building material, but few people dispositionally believe that 33,873×0.81=27437.13.

It's worth noting that this criterion—that genuine beliefs are poised for action and inference—provides a principled way of avoiding what is sometimes known as "cognitive bloat" in the literature on the extended mind (Rupert 2004; Sprevak 2009; Allen-Hermanson 2013). According to the extended mind hypothesis, minds may literally be partially constituted by external resources, just in case these external resources play a functional role in cognition that would qualify them for being part of the mind were they internal. The cognitive bloat objection is that the extended mind hypothesis commits us to thinking that all external informational resources could be literally part of the mind, if we're hooked up to them appropriately. If Otto knows where the Museum of Modern Art is, in virtue of the information he has written down in his notebook, then mightn't he also know everything in Wikipedia? That's the problem of cognitive bloat. We think our fifth condition avoids the problem. If—implausibly—Otto is indeed set up such that all the information in Wikipedia is poised for action and inference, then he *does* know all of its content. In realistic scenarios, however, the information is only effortfully available to him, and he does not believe it unless and until he retrieves it.

Since we're concerned here with autonomy, we want to pause to examine a potential objection that turns on it. One might think that Van Leeuwen's criteria for belief (with or without our added fifth condition) are belied by the very possibility of nonautonomous belief. A nonautonomous belief might, after all, be brought about by indoctrination or brainwashing. Such a belief is not the product of evidence. Yet, it is still a belief. We agree that, in principle at least, beliefs could be inculcated via methods that have nothing to do with evidence. However, a genuine belief is constrained by evidence in the form of the agent's other beliefs (Dennett 1975) and is responsive to evidence once formed (Helton 2018).[4]

Although we are not persuaded by Van Leeuwen's case that religious attitudes are rarely believed, we agree that some such attitude are unlikely to qualify. We think that the central (professed) religious beliefs of the faithful are genuine beliefs. Christians, Muslims, and Jews genuinely believe that God exists, that God created the world, and so on. The divergences from the dispositional stereotype (Schwitzgebel 2002) of genuine belief Van Leeuwen points to are better explained by the deferential content of the belief than by the attitude the person takes to it (Levy 2024b). But the faithful likely don't believe every religious assertion they might sincerely utter. Some they mouth as empty platitudes, giving little thought to them. Perhaps they even take themselves to believe some that are better characterized as a different sort of attitude altogether.

Van Leeuwen's criteria allow us to plausibly distinguish beliefs from other states—from what he calls credences, on the one hand, and from what he calls other secondary attitudes, on the other (wishing that *p*, hoping that *p*, desiring that *p*, and so on). Again, there may be a case for allowing the extension of the term "belief" to range more widely. We don't intend to show that "belief" *must* be reserved for states satisfying his criteria to a sufficient degree. (We expect some states will fall at the borders of belief, although we—unlike Van Leeuwen—think that the criteria form a homeostatic cluster: States that have one of these properties to a high degree will tend to have them all.) Our aim, rather, is to delineate a class of propositional attitudes that are paradigmatically beliefs and assess the extent to which groups may be said to have states like *that*. Just as there are belief-like states that individuals may mistake for beliefs, we expect that there are belief-like states that have been mistaken for group beliefs (perhaps there are group shmeliefs, for example), and just as there may be states of individuals that are neither clearly beliefs nor clearly *not* beliefs, so there may be states of groups that are neither clearly beliefs not clearly *not* beliefs. None of that is our concern: We're concerned only with states that have all the marks of beliefs to a high degree.

If there are group beliefs in this strong sense, they will satisfy Van Leeuwen's criteria *at the group level*. A group wouldn't count as having a belief if it was true that it possessed states that were, say, evidentially vulnerable or non-compartmentalized (only) when the individuals who constitute it confront evidence or engage in inference. Rather, it must be the case that these states govern the *group's* inferences and its behavior. The belief will govern the group's behavior broadly, not just in certain contexts. It will respond to evidence that the group has, and it will be involuntarily acquired; voluntariness will affect only inquiry, not the resulting belief.

With our target in mind, let's turn now to existing proposals for group beliefs, and their limits.

3.2 Distributed Non-Summativism

It should be plain that some extant proposals for group beliefs are simply nonstarters when measured against our criteria. Consider joint acceptance accounts. On a joint acceptance account, a group believes that *p* if and only if (enough of) its members jointly accept that *p*, conditional on the other members accepting that *p* and acting accordingly (M. Gilbert 1992), or if its "operative members"—those with the authority to decide what the group believes—jointly accept that *p* (Tuomela 1992). There's an obvious objection: An acceptance is not a belief (L. J. Cohen 1995). Acceptances fail every one of Van Leeuwen's criteria. They may be voluntarily acquired: An agent decides to accept a proposition in view of some practical aim (Hakli 2006). Acceptances are compartmentalized: The agent who accepts that *p* (but does not believe it) adopts a policy of guiding behavior and inference by reference to it in settings in which achieving a particular practical aim is relevant. (Admittedly, the truth or otherwise of *p* may simply never arise outside such settings: Consider some scientific claim like "such and such a hypothesis is true." Nevertheless, beliefs may be distinguished from acceptances, inter alia, by reference to counterfactuals.) Acceptances are not sensitive to evidence—they may even be adopted in the teeth of evidence (consider the agent who decides to accept that her friend has her best interests at heart, despite the evidence).

Perhaps proponents of joint acceptance accounts of group belief might argue that the terminological identity between individual and group acceptances is misleading. Perhaps, that is, the states of groups that are brought about by joint acceptance by individuals have quite different, and more belief-like, functional properties than individual acceptances. Be that as it may, joint acceptances must fail a central test for being a belief. There are no evidential constraints on joint acceptance, and joint acceptance can occur for any reason whatsoever, even in the teeth of the evidence. Although we've conceded that genuine beliefs could be brought about in ways that bypass the need for evidence, they cannot persist in the face of evidence. Joint acceptances can. Unsurprisingly, then, they are devastatingly vulnerable to Jennifer Lackey's (2021) criticism that joint acceptance accounts cannot plausibly account for group lies.[5] That they should have this inability follows from the fact that joint acceptance is not constrained by evidence.

Of course, it's open to a defender of the joint acceptance account of group belief to argue that group belief is different from individual belief. They might concede, that is, that joint acceptances do not satisfy the functional criteria that are plausible conditions for individual belief, but maintain that they are nevertheless the best candidates for group belief (Wray 2003). We think this is false, however. We think that beliefs that satisfy Van Leeuwen's criteria may be ascribed to (some) groups.

A group belief (in this demanding sense of "belief") will be involuntarily adopted by the group in light of the evidence available to it or—at very minimum—will fail to persist in the face of evidence that conflicts with it. That is, the belief will aim at truth. Aiming at truth follows from being constrained by the evidence.[6] We think that group beliefs, in this demanding sense, must be postulated in order to make sense of group lies. If Philip Morris lied about whether tobacco causes cancer (to use Lackey's central example), it's because the group has a belief that conflicted with its public assertions.

The natural way to secure these properties for group belief is by adopting a summativist view. A group belief that is attributed in virtue of the beliefs of its members might inherit from the latter the properties that make the state a genuine belief. It will have mind-to-world direction of fit, it will be involuntary, it will be evidence responsive. As Lackey (2021, 53) puts the thought here, "group belief needs to be anchored by individual beliefs" if group belief is not to be voluntary. It is these considerations that motivates her group agent account, which requires that a significant percentage of group members believe that *p* for the group to believe that *p*.[7]

We think the claim that group belief needs to be anchored by individual beliefs is mistaken. Although summativism is *one* way of securing the properties distinctive of belief for group beliefs, it's not the only way. Distributed models (Bird 2019) may do so as well. At least, our distributed model can do so, we argue.[8]

If a group adopts a procedure for adopting beliefs, and it adopts that procedure because it is evidence-sensitive and aims at the truth, then the resulting state would have a mind-to-world direction of fit and would be involuntary (involuntary, that is, given that the group has voluntarily adopted the procedure). We think that this is exactly what the groups that are the fundamental units of scientific research actually do.

Here's an example to make the discussion just a little less abstract. Suppose a laboratory aims to test the hypothesis that support for policy *a* is partially caused by (apparently unrelated) disposition *b*—say, that support for deposit schemes on glass bottles is partially caused by a heightened sense of one's mortality. Typically, the lab aims to test that hypothesis because some of the group members think it's plausible,[9] but that's not always the case. Testing the hypothesis may be required as a condition of publishing some other result, for example: Reviewers may require such testing to rule out alternative explanations of a finding. Alternatively, lab leaders may simply be interested in the claim and agnostic about whether it's true, or (as in one case we know of) may have inherited leadership of the group and be actively hostile to the hypothesis. The other members of the lab (or labs as is increasingly the case) may have little interest in the hypothesis specifically or may be skeptical whether it's true. Nevertheless, they play their part in testing it.

In much of contemporary science, tasks are divided between lab members in complex and overlapping ways. Not only may lab members have a diversity of attitudes toward the hypothesis under test, they also need not have any solid grasp of how their particular task contributes to testing the truth of the hypothesis. Because tasks are distributed widely across different fields of expertise, the people working on one task may have little knowledge of—or interest in—the broader research question. Of course, there must be at least one person who believes that an individual member's contribution bears on the question, but that person or those people may lack the expertise to do the task themselves: They delegate it instead (Bird 2010). Members of the subgroup need not even believe *that* their task contributes to testing the truth of the hypothesis (perhaps the lab has decided to use Bayesian hypothesis testing, rather than the traditional null hypothesis significance testing, and the lab member tasked with the analysis is skeptical of the approach).

In contemporary science, answering a research question might therefore require work distributed across multiple people, in multiple labs, and with multiple kinds of expertise, and only some of the individuals involved need have any attitude at all to the question. Although the tasks are assigned because someone with the authority to delegate believes that collecting data on *that* population, or analyzing data in *this* sort of way, or even developing an algorithm of this sort or building a physical tool like that would contribute to answering the question, there need be no individual who believes or even understands how each component contributes to the whole. Tasks may be completed independently. Not rarely, some tasks are entirely independent of one another, some are interdependent, with reciprocal linkages between them, and yet others produce data or findings that serve as inputs into other tasks. When all tasks are completed, those group members who have been assigned *this* task integrate them into a bigger picture.

In the simplest sorts of cases, a positive outcome on each task provides evidence for the hypothesis *h*. In other cases, some tasks produce an output that cannot, on its own, distinguish between *h* and some rival but provides evidence in support of *h* when combined with other findings. In none of these cases need those who perform the tasks have any attitude in particular toward *h*. Nevertheless, if the experiment is appropriately designed (by operative members of the group, typically—although an experiment might be designed by people outside the group or by an AI for that matter), they contribute toward testing it. The resulting paper, combining and coordinating the results of the distributed labor, may then be written, concluding that mortality salience mediates support for bottle recycling programs or what have you. The proportion of group members who believe the conclusion the group argues for (and, we suggest, believes it *qua* group) might be tiny. At the limit, it might be zero (see Dang and Bright [2021] for an argument that belief is not a norm of assertion for scientific claims).[10]

Our account is non-summativist. But it satisfies all the desiderata provided by Van Leeuwen's criteria as well as our poise condition. The resulting group belief is involuntary. The group believes that *p* when and because the integration of the tasks that were engaged in to test whether *p* yields the result: *p*. Group members need not believe the claim; the *group* believes it because its procedures commit it to it. Just as obviously, the belief that *p* is evidentially vulnerable. The group would abandon the belief were it to be shown that they had made a mistake in processing the data, or an alternative hypothesis were proposed that was a better explanation of the data, or if another group produced overwhelming evidence (perhaps from a far better powered study, with more impressive evidence) that the belief were false. The belief that *p* has a mind-to-world direction of fit: The procedures were followed because they were held to bear (better than any alternative available to the researchers) on whether *p*. Finally, the belief is poised: In fact, the group has established procedures for outputting the belief as the group's findings.

Because our account requires that beliefs be produced by evidence-sensitive procedures, it explains the possibility of group lies and bullshit (sadly, both exist in scientific publishing). Suppose the group is funded by Philip Morris International to conduct research designed to cast doubt on the claim that smoking tobacco causes cancer (in fact, tobacco companies spent millions funding research, some of it well conducted, to precisely that end; see Oreskes and Conway 2011), but the findings actually support the mainstream view that tobacco is carcinogenic. If the group reports that their research shows that tobacco is not carcinogenic, it lies. It lies even if the group leaders actually believe that tobacco is not carcinogenic. They have competently followed a set of procedures, the results of which entail that the group believes tobacco is carcinogenic, so their assertion is a lie. It's easy to tweak this sort of example so that the utterance is made without regard for its truth and is therefore bullshit.

We think there's a plausible case to be made that a group can lie even in cases when it has competently followed a procedure that entails the belief it reports. This is the case because there are constraints on what procedures a group may adopt—constraints that are entailed by the requirement that they aim at truth. The group must adopt the best set of procedures it takes to be available. Many of the actual studies that supported the tobacco industry's line were well conducted (in many ways) and produced genuine evidence that cancers arose mainly from (actually rare) genetic predispositions. But they were deliberately underpowered in the knowledge that small studies are more likely to produce unreliable findings (O'Connor and Weatherall 2019). We're comfortable saying that the groups that published these findings lied.

It's easy to see that, on our account, group beliefs satisfy the evidential vulnerability and non-voluntariness condition. It's harder to assess whether they satisfy Van Leeuwen's two other criteria: no compartmentalization and cognitive governance. We think it's plausible to think that they do. The belief attributed to a scientific research group is compartmentalized in a sense: It is restricted to scientific contexts. But that restriction is simply a function of the contexts in which the research group *exists*. The belief doesn't guide taking the bus or donating to charity, for example, because scientific research groups do not engage in these behaviors *qua* research groups. It guides the inference and the behavior of the research group in those contexts in which it exists and functions as a scientific research group.

On our account, beliefs may be appropriately attributed to groups when their assertions are governed by procedures that are well designed to be sensitive to evidence, and only to evidence, whether or not the members of the group believe these assertions.[11] This is a non-summativist account, of course. Our defending it, however, does not commit us to rejecting summativism. We think a disjunctivist account, on which group beliefs are veridically attributed in virtue of distributed processes that respond to evidence *or* on the basis of the beliefs of a majority of its members, has *something* going for it, although it also faces challenges.

A fully developed disjunctivism of the sort we envisage here must do more than set out conditions for each disjunct, of course. It must also develop principles for adjudicating cases in which the disjuncts give conflicting verdicts. Suppose a research group deploys appropriately truth-tracking processes and thereby comes to believe that *p*, while a significant majority of its members hold a different view. Perhaps a disjunctivist might bite the bullet and live with the inconsistency: attributing conflicting beliefs to the group in this sort of case (some philosophers have been willing to attribute inconsistent beliefs to individuals; see, e.g., Bendaña and Mandelbaum 2021). But any view that attributes conflicting beliefs to one and the same individual or group must engage in fancy footwork of some kind or another because *p* and not-*p* cannot simultaneously satisfy our conditions for belief. At minimum, they can't both exhibit cognitive governance and poise.

So, the disjunctivist must draw some distinctions. They may say, for example, that *qua* scientific research group, the group believes that *p*; *qua* group of scientists, the group believes that ~*p*. In such a case, an individual who purported to speak for the group would need to delineate in what capacity they spoke.

3.3 Bird's Distributed Model

The model we've been defending is, as we mentioned, a "distributed account," in Bird's terminology. It's therefore important we compare it with Bird's own account (Bird 2014).

It must be noted at the outset that we suspect that Bird's account of group knowing is not aimed at addressing the same question that we, and most other contributors to the debate on group belief, are concerned with. We're concerned with the conditions under which a group of agents may veridically have a belief attributed to it. Bird seems to have in mind the conditions under which an abstract entity, such as science, counts as knowing something. As he says, he's concerned not with groups that are (even loosely) organized but rather with "wider science" (Bird 2014, 48) as well as with such collectives as nations and even humanity. We aim to explain how, for example, it might be true that a group that produces a paper for publication knows the main findings of that paper. Bird is concerned with the conditions under which it is true that, for example, we know that smoking causes cancer or that North Korea knows how to build a nuclear weapon (Bird 2014, 52). Although—as these locutions show—it is natural to talk about knowledge in this sense, we're not sure that this sense of knowledge entails mental states, or even functional equivalents of mental states, at all. As we see it, Bird is concerned not with whether any group actually has such states but rather with whether knowledge is available to the group.

If Bird does have the same target as we do, then we reject his account because it fails our poise condition. Bird does set down a condition something like this one for group knowledge, but it is much weaker than ours—too weak, we think, to secure belief. It is, he says, enough if the knowledge is accessible "in principle," where for knowledge to be accessible in principle is for it to be stored somewhere such that if someone with the capacity to understand it were to consult the right archive (or what have you), they would be able occurrently to know it (Bird 2014, 48). Bird gives the example of Dr. N, whose paper is published in a neglected field and who then dies, as does everyone else who has read the paper. Bird argues that Dr. N's discovery is "part of scientific knowledge," even once all N's readers have died and the result is forgotten (Bird 2014, 48). Bird argues that N's discovery is accessible; it's not as though it is locked up in a drawer (as in the case of Dr. Q, who features in another of Bird's cases).

It's apparent, though, that Dr. N's discovery is very far from poised for action and inference. Those working on problems on which N's work bears cannot use it to support their arguments, for example. It is accessible only via a route that is highly effortful. We think this failure is diagnostic: Most likely, Bird does not have the same target in mind as do we and, we think, as most of the other contributors to the debate over group belief do. There is a perfectly good sense in which N's discovery is "part of scientific knowledge," we agree, but that sense does not require us to attribute a belief to any group. This is a sense of knowledge that does not entail belief or indeed any mental state at all. It is the same sense in which libraries contain knowledge.[12]

There's another route to the same conclusion: Bird's explicitly Durkheimian account of the nature of groups allows for the existence of groups that do not possess beliefs at all. Durkheim argued that groups may cohere on two different bases: mechanical and organic solidarity. Mechanical solidarity is achieved via shared beliefs. Groups unified on this basis are unified because the individuals who compose them are similar in important respects. Organic solidarity does not require similarity. Rather, it requires that individuals be functionally integrated (organic unity, here, is the kind of unity supposed to be characteristic of organisms).

Bird's account of the basis on which groups cohere allows for very loose links between members. On his Durkheimian account, organic solidarity is brought about by functional integration. Bird argues that the mutual dependence that arises from the division of labor, in science and in society, is sufficient for organic unity. We think that groups, in this undemanding sense, need not be sufficiently unified to possess beliefs. For a group to possess beliefs, the attitudes attributed must play an appropriate role in the action and inference of the group. It's clear this condition need not be, and regularly is not, satisfied by contemporary nations. In Bird's sense, Russia believes that climate change is a serious problem for the world. But Russia's behavior makes it wildly implausible that Russia believes that. The knowledge is accessible in Russia (there's nothing wrong with Russian science), but the state does not play the role characteristic of belief.

For a group to possess beliefs, it's not enough that there is some principle of unity binding members together: The beliefs must be appropriately integrated into action and inference control mechanisms. The groups of researchers to whom we think beliefs are appropriately attributed are aware of themselves as constituting groups and as contributing to a common project. They have action and inference mechanisms that take their beliefs as inputs. Their beliefs satisfy the no compartmentalization, cognitive governance, and poise conditions, as well as the evidential vulnerability condition. Bird's account secures only the last.

None of this is to criticize Bird! His account is the best on the market of the target *he* has in mind, and that target is one well worth theorizing about. Moreover, his work on scientific groups is illuminating for our purposes, as well as his, since he often discusses cases involving groups that are much more tightly integrated than his Durkheimian exemplars: integrated, in particular, by procedures for producing, processing, collating, and publishing information. We've drawn from his discussions of these more integrated groups in developing our own distributed model.

3.4 Group Beliefs and Shmeliefs

In the first chapter, we distinguished genuine belief from some impostors—belief reports that are insincere, expressive, or sincere but mistaken. It is, we argue, a virtue of the criteria we've adopted and adapted from Van Leeuwen that it allows us to distinguish beliefs from these doubles. These impostors fall short because they fail to be evidence responsive, are compartmentalized, or are acquired voluntarily. Let's return now to such cases and a question we raised but did not answer. Might there by group *shmeliefs*? Perhaps even more interestingly, might genuine group beliefs be constituted or realized by individual attitudes like shmeliefs?

Shmelievers are people who take themselves to believe something but are mistaken. They are mistaken not simply for the kinds of reasons we all make such mistakes from time to time: We haven't "put two and two together" or traced out all the implications of our beliefs. One of us recalls being challenged by a well-known philosopher on proclaiming that they believed some contentious philosophical proposition: "You don't believe that." That philosopher pointed to some of our other commitments and noted their inconsistency with the claimed belief. He was right. Although the belief report was mistaken, though, it wasn't a shmelief. The shmeliever sticks to their guns in the face of the apparent evidence against their belief, explaining away the inconsistency or even claiming to reject some extremely plausible and largely noncontentious proposition to maintain their shmelief. Of course, we want to leave room for bizarre beliefs: Some people really believe some really odd things.[13] A bizarre belief satisfies our conditions for being a belief; a shmelief does not.

We think, for example, that many of the sorts of beliefs that are characterized (sometimes erroneously) as conspiracy theories are shmeliefs. One giveaway is that the person may cite as evidence for their belief considerations they cannot possibly take really to make it more probable. For example, they may cite coincidences as evidence. Consider the widely circulated factoid that "delta omicron" is an anagram of "media control" (see Levy [2022a] for further discussion). Those who cite factoids like these are almost always broadly rational agents. Nothing in their lives indicates that they possess personal epistemic standards or procedures that would bring them to regard these factoids as evidence. Of course, they might be insincere (some surely are). But a significant number of them will go on subsequently to reject the theory and to characterize their former selves as believers. We think that these agents are right about their sincerity but wrong in how they characterize their former selves. It's almost impossible from the inside to distinguish shmelievers from believers. In any case, we proceed on the assumption that there are individuals who are shmelievers.

It's plain that summativism cannot allow for group beliefs to be realized by individual shmeliefs. On a summativist account, if the base attitudes are not beliefs, then any group attitude constituted or realized by them is itself not a belief. What such attitudes might be (if they exist) is another matter. Perhaps we should be summativists about shmeliefs: Group shmeliefs might be constituted by individual shmeliefs. Perhaps summativism is true of other non-doxastic attitudes (imaginings, hopings, and so forth). We won't explore the question, but we find this sort of summativism at least somewhat plausible. After all, it's news to no one that conspiracy theories and the like are *social* phenomena. Indeed, on some accounts, they have as a principal function the signaling of commitment to a group (Mercier 2020; Funkhouser 2022; Ganapini 2023).

Joint acceptance accounts, on the other hand, might allow for group beliefs that supervene on individual shmeliefs. This is especially likely to occur in cases of pluralistic ignorance (Miller and McFarland 1991). In such cases, group members privately believe that $\sim p$ but avow that p because they believe that everyone else believes that p and they fear ostracism or alienation from the group.[14] In this sort of case, they may jointly accept that *p*, conditional on everyone else doing so, and therefore bring it about that their group believes that *p*. This is an unsurprising result: Since joint acceptance accounts readily countenance cases in which the group believes that *p* while few or no members believe that *p*, it is unsurprising that they allow for group beliefs in this sort of case. We see no reason why the same mechanism mightn't work with shmeliefs. Conspiracy theorists might not believe that the moon landing was faked, for example, but they might readily jointly accept this proposition, conditional on their fellow group members doing the same.

Finally, let's consider distributed accounts, like the one we've defended above. Like joint acceptance accounts, our account allows for group beliefs in cases where few or no members have a belief with the same content. We think this kind of case can allow for group beliefs where individuals have a non-beliefy attitude to the same proposition. Shmeliefs, for example, may have any content at all: Nothing excludes the members of a research group shmelieving the product of their research group. Perhaps such cases are actual. Suppose that the research leads to a very counterintuitive conclusion (as scientific research, notoriously, sometimes does). In that case, group members might endorse the product of their research without being able to draw inferences or govern their behavior on its basis.[15]

In the case we just imagined, though, the group genuinely believes that *p* because it is committed to *p* in virtue of its use of reliable and evidence-sensitive procedures. Obviously, such cases are very different from the sort of conspiracy theories with which we began the discussion above. There's no content-based restriction on which propositions might qualify for shmeliefs, so we don't rule out shmelievers constituting a believing group, but ideological shmelievers of the sort we sketched will not. As we saw, they betray the fact that they are shmelievers by using evidence that is wildly unreliable. Our non-summativism will therefore give a different verdict on these cases to joint acceptance accounts, which place no constraint on the evidence a group uses.

On a joint acceptance account of individual shmelief coupled with group belief, the causal arrow runs from shmelief to belief: Group members jointly accept that *p* because they shmelieve that *p*. On our model, the causal arrow more often runs in the reverse direction, with group members shmelieving that *p* because the group believes that *p*. In such cases, in effect, the group-level belief serves as testimony on the basis of which they try to believe that *p*, failing due to the counterintuitive content of the belief (or its conflict with implicit biases or whatever). In cases like this, group membership is incidental: An outsider might shmelieve that *p* on exactly the same basis as a group member.

3.5 Can Group Belief Be Heteronomous?

It won't have escaped the reader's attention that although this discussion may be interesting (we flatter ourselves that it is), we have not yet discussed the ostensible topic of the book. It's all very well to defend an account of group belief and to set out the options for other group attitudes, but this discussion must be preparatory to the development of tools for assessing the *autonomy* of such states.

To our knowledge, no one has ever inquired into whether group beliefs can be qualified as autonomous or heteronomous. One possible explanation for this lack of discussion is that group epistemic autonomy might be thought to be a trivial question, either because it doesn't matter to us whether the groups to which we belong are epistemically autonomous, or because groups are just not the sort of thing that can be autonomous in this sort of way.

We think the epistemic autonomy of groups can be an important question nevertheless. It's certainly true that groups enjoy a bad reputation as epistemic agents. "Groupthink" more or less *means* nonautonomous thinking, and groups are routinely held to be less intelligent than the individuals composing them (Terry Pratchett: "The intelligence of that creature known as a crowd is the square root of the number of people in it"). Mercier and colleagues (2015) report that many of their participants expected groups to do worse on a cognitive task (the Wason selection task) than the individuals who composed them.

Although the lack of previous inquiry into group epistemic autonomy may stem from the belief that groups just can't be autonomous because they are too subject to nonrational pressures, that belief is mistaken. In the study just mentioned, success on the Wason selection task rises from around 10 percent (in the individual-alone condition) to 70 percent in the small-group condition, and such epistemic success is not unusual. Under a wide variety of quite undemanding conditions, groups are more reliable epistemic agents than individuals can be. The success of science is dependent on this fact.

We think, therefore, it's very much a live question whether a group's beliefs are autonomous. Although the conditions under which groups do better than individuals in deliberation are quite undemanding, they can be flouted, deliberately or not. Groups do worse when they suffer from the problem of "hidden profiles," which occurs when group members withhold information (Sunstein and Hastie 2008). This may occur from the premature pooling of information or of opinions. Groups may, for example, be subject to information cascades, which can occur when agents express their opinions sequentially. In such cases, a group belief can emerge shockingly early, leading those agents later in the sequence to withhold their private information. The result may be a group that believes that *p*, composed of individuals most of whom possess information that pooled strongly indicates or even entails that not-*p* (L. R. Anderson and Holt 1997). The same situation can arise via the simpler route of deference to perceived authorities, of course (Sunstein 2006).

Hidden profiles are avoided by structuring groups and group knowledge production to incentivize or require the sharing of information. Modern science is dependent on such structures: The peer review process (*deeply* imperfect as it is) helps to mitigate the problem of hidden profiles by making it possible for junior researchers to take issue with the claims of their seniors in the review process without fear of retaliation.[16] Post-publication review takes on an ever more important role in this regard, allowing the exposure of scientific claims to scrutiny from a very wide variety of perspectives, including the perspectives of those who do not have much to risk from dissent. In any case, we think there's a plausible case to be made for holding that at least some of the ways in which groups go astray in forming beliefs involve the formation of heteronomous, or at any rate nonautonomous, belief.

This is especially obvious in the case of group belief that arises through premature deference to perceived authority. Such deference, in the individual case, is *paradigmatic* of heteronomous belief. Those who believe that *p* only because some epistemic authority says so *and* who possess evidence that indicates that not-*p* are paradigms of heteronomous believers. We see no reason to think that the same may not be true with regards to groups. Moreover, we're confident that groupthink is a real phenomenon, even if it occurs much less often than we're culturally disposed to think. Take the phenomenon of "herding" in opinion polls. Herding occurs when opinion polls adjust their weightings in the light of the findings of *other* opinion polls. Such adjustments are, to some degree, defensible and sometimes even obligatory. If a poll taken tomorrow in the United Kingdom indicated overwhelming support for the Green Party, we'd be pleased but very skeptical. Given that the level of support for the Greens is known to be relatively small, and nothing has recently happened to explain a surge in their popularity, the best explanation would be polling error: Perhaps the pollsters inadvertently massively oversampled meetings of party supporters—hence the need to adjust weightings to bring the poll more into line with known realities.[17]

This isn't yet herding. Herding occurs when the opinion polls are excessively deferential to one another, such that they lose touch with the actual rates of support of political parties in the community. Polls may fail to shift in the face of actual population-level shifts in support for parties or policies due to herding. We think this qualifies as groupthink. In the light of these considerations, we think it matters how a group belief arises. Just like individuals, groups value—or should value—*thinking for themselves*. A group thinks for itself when its cognition is governed by the procedures it has imposed on itself and it has selected these procedures because they are judged to be evidence sensitive and reliable.

It is in this light that we now turn to the autonomy of group belief. We aim, in particular, to assess the prospects for non-summativism *about autonomy*. A non-summativist account of group belief holds that groups may have genuine beliefs that dissociate from the beliefs of its members. (Indeed, these accounts allow for a double dissociation: Groups may have beliefs that dissociate from the beliefs of their members, and groups may lack beliefs held by most or even all the individuals who constitute them.) A non-summativist account of group belief *autonomy* holds that groups might autonomously believe that *p*, even in cases in which members *nonautonomously* believe that *p* (or, perhaps, nonautonomously believe that not-*p*; in such a case, we would have dissociation at the level of the content *and* at the level of its autonomy). On a summativist account, the degree of autonomy of a group belief cannot come apart from the degree of autonomy of the beliefs of the individuals who compose the group. Summativism about group belief need not entail summativism about group belief autonomy, we will suggest.

Below, we consider the options. We won't assume any particular account of group beliefs in this section; rather, we want to map the territory. For ease of exposition, we will use the terms "summativism" and "non-summativism" without further qualification only to refer to accounts of group belief rather than group belief autonomy.

3.5.1 Summativism

On a simple summativist account, a group, *g*, believes that *p* just in case a significant enough proportion of its members believe that *p*. At first glance, it's hard to see how summativism about group belief can come apart from summativism about group belief autonomy. Surely, if the members of the group who bring it about that the group believes that *p* possess their belief autonomously, then the group belief must itself be autonomous (assuming that autonomy is a property that group beliefs can have at all)? We think there are grounds for thinking that the autonomy of group-level and individual-level beliefs can dissociate on a summativist view.

Here's one way in which autonomous group beliefs might be constituted by heteronomous individual-level beliefs on a summativist account of group belief. Suppose *g* believes that *p* in virtue of the fact that a majority of its members believe that *p*, but that many of those members believe that *p* because they have been oppressively socialized into that belief. Such socialization is widely held to be autonomy subverting, even when those who are subject to it would endorse it on reflection (Stoljar 2014; Oshana 2006). But it is far from obvious that *group* endorsement of the belief and the processes that sustain it is not sufficient for group autonomy. In this sort of case, individual-level beliefs are very plausibly heteronomous, in part because the endorsement is *itself* heteronomous, but the group may hold the same belief autonomously. One reason we think it is (at minimum) an open question whether the group's belief is autonomous is because it's hard to make sense of mechanisms of oppressive socialization that operate at the group level. Groups, and their representatives, are the *agents* of socialization, and the group and its norms are what agents are socialized into. We're somewhat skeptical that it makes much sense to think of such processes as operating on the group itself.[18]

What about the converse sort of case, where individual-level autonomous belief constitutes group-level heteronomous belief? We think such cases are also possible.

Consider a case in which individual members' beliefs that *p* are stabilized by the knowledge that the group believes that *p*, and if individual members did not believe *that*, they would not believe that *p*. Suppose *p* is "I should send my children to a private school." Members of *g* might hold this belief, rationally, because they believe that the fact that my group believes that *p* ensures underinvestment in government schools. Many or most members of *g* might believe that *p* only because the group believes that *p*. Were fewer group members to believe that *p*, the evidence for *p* would be defeated or at least weakened, and agents would be less likely to believe it (which would further weaken the evidence for it, leading to the rapid loss of the group belief).

Now, what should we say about the autonomy of the beliefs, both those of the individual members who constitute the majority or plurality on which the group belief depends and those of the group belief itself? We think this is plausibly described as a case in which heteronomous group belief is constituted by autonomous individual belief. We cannot see grounds for holding that the individual-level beliefs are heteronomous. Granted, they respond to a social fact that the individual would prefer to be false, but they do so rationally. There's no reason to think a belief is heteronomous in virtue of being a response to such a lamentable fact. Lots of our beliefs are formed in response to lamentable facts: That does not render them heteronomous.

Perhaps it might be thought that the history of the belief's acquisition by individuals renders it nonautonomous. We don't think that history need be autonomy subverting. That history may involve highly regrettable processes, whereby individuals' reluctant acceptance that *p* (or the increase in the credence they assign to *p*) constitutes evidence for the very belief they want to be false. But this history is not a history of manipulation or of the bypassing of rationality. Again, there seems no good reason to think individual-level beliefs are heteronomous in virtue of this history.

Nevertheless, we think there might be grounds for holding the group-level belief to be heteronomous. Once we recognize the possibility of group-level autonomy dissociating from individual-level autonomy (as we think we should, in response to the oppressive socialization case), the way is open to ask the questions independently, and we cannot take the fact that individual belief is autonomous to settle the issue. We must ask, instead, about the *group-level* epistemic properties of the belief. Does it reflect the group thinking for itself? Although the belief might be the product of individuals thinking for themselves—that is, responding appropriately to their evidence—the group-level belief is *not* a response to evidence. It is stable not because it responds to evidence, but only because it is self-sustaining. Although self-sustaining beliefs may not always be heteronomous (think of an exuberant optimism, which might enable the production of evidence that supports it), an unwanted self-sustaining group belief does seem to be resistant to reflection in a way that threatens its autonomy. The group-level belief is an unwanted belief that is held because its being held brings about the evidence that sustains it. This fact renders it resistant to revision in a way that we think makes its autonomy questionable.

To better see what we have in mind, it's worth contrasting the individual-level belief that *p* with the group-level belief that *p*. Although it's true for each individual that *their* believing that *p* is part of the total set of facts that makes it rational to believe that *p*, their belief that *p* is very far from making the difference in this regard. Their failure to believe that *p*, or believing that not-*p*, would not make a discernible difference to the set of facts to which the belief responds. In contrast, a change in the *group's* belief would be sufficient all by itself to alter what it would be rational to believe. The group can consistently believe that *p*, that it would be (non-epistemically) better to believe that not-*p*, and that there is no reason to believe that *p* independent of the fact that the group believes that *p*. Since the belief is unwanted, and it would be rational to abandon it just in case it was abandoned, the inability of the group to abandon it (for want of any mechanism to do so) seems to us to threaten its autonomy. Again, it does not reflect the group's rational inquiry into whether *p*.

We think, therefore, that summativism about group belief autonomy permits double dissociations between individual- and group-level autonomy. It's worth noting another wrinkle: Summativism might imply heteronomy on the part of agents who do not share the group belief but who nevertheless belong to the group. Suppose the group believes that *p* because a majority of its members believe that *p*. At least under many conditions, the minority who believes that ~*p* cannot disavow group membership. They belong to a group that, they might recognize, believes that *p* and acts on that belief while themselves rejecting *p*. They may well feel deeply alienated from their group in virtue of this divergence (think of the egalitarian who belongs to a racist group). Although they do not share the belief, *qua* group members, it is nevertheless in some sense theirs. Although the group-level belief might be autonomous, it might be appropriate in such cases to ascribe the heteronomous belief that *p* to a subgroup composed of individuals who believe, *qua* individuals, that not-*p*.

3.5.2 Joint Acceptance Accounts

On joint acceptance accounts, *g* believes that *p* just in case its members (or its operative members) jointly accept that *p*, where to jointly accept that *p* is to accept that *p* conditional on other members of *g* accepting that *p*.[19] We think that such accounts allow for a dissociation between individual- and group-level autonomy.

There's a straightforward route from individual-level autonomous belief to group-level heteronomy on joint acceptance accounts. It's supposed to be a feature of such accounts that group-level belief is unconstrained by evidence. The board that jointly accepts that *p* might do so even while knowing that the evidence tells against *p*. Now think of many paradigmatic cases of heteronomous belief formation. It's no threat to our autonomy that a belief is forced upon us by the evidence. But it is a threat to autonomy if a belief is brought about through processes that bypass rational cognition (this is a central contention of much of the literature on nudges, after all). We think this is a (strong) reason to think that the group's belief is heteronomous.

This may seem counterintuitive: After all, the belief was adopted by the free and uncoerced actions of the group members. But we're concerned with the autonomy of the *group's* beliefs, not those of its members. Here's an analogy. An agent's cognitive states are realized by subpersonal processes and mechanisms, but the autonomy of their states is a personal-level property of theirs. The epistemic properties of subpersonal states may be very different from the epistemic properties of personal-level states. Groups stand to individual members in something like the sort of way in which personal-level processes and states stand to subpersonal processes and states: The properties of the second are often not inherited by the first. The fact that individual members are autonomous in how they bring about the group belief therefore need not entail that the group belief is itself autonomous.

Consider a case in which a group has instituted processes for gathering evidence and weighing it, with a view to forming group-level beliefs. Such a group aims at truth; it aims to accept beliefs only when these beliefs are true. Since (we're supposing) joint acceptance is sufficient for group level belief, however, its beliefs are unconstrained by these processes. In the case we're envisaging, the group accepts that *p* while knowing that *p* would not pass muster were it subjected to the deliberative processes it has instituted. Here, belief formation bypasses the *group's* rational deliberate processes. We think this fact constitutes grounds for thinking the belief to be heteronomous.

Joint acceptance accounts also allow for the possibility of autonomous group belief with the same content as heteronomous member belief. Suppose members of *g* have been oppressively socialized into believing that *p* and that their beliefs are heteronomous in virtue of that fact. Nevertheless, they might jointly accept that *p*. Perhaps they might even be autonomous *in* doing so: They may be making the best of a bad situation. It might be better that their group believes that *p* because individual members are better able to achieve their goals, given their heteronomous beliefs, if the group also believes that *p*. In this sort of case, the autonomous group-level belief has the same content as the heteronomous individual-level beliefs of group members.

Finally, we suggest, joint acceptance accounts may give rise to individual-level heteronomy, but do so without threatening the autonomy of the *beliefs* of the individuals. Suppose that every member of the jury believes that the defendant is guilty, but they jointly accept that they're not guilty. They may recognize that the defendant was rightly acquitted but remain acutely aware that justice has not been done (perhaps the defendant is getting away with an awful crime). They could be expected to feel alienated from the group-level belief and unable to identify with it in the way required by at least some accounts of autonomy (e.g., Frankfurt-inspired accounts that require that the agent be satisfied with their own beliefs). We don't think this case is well characterized as involving heteronomy at the group level. Here, the individual-level belief (that the defendant is guilty) might be autonomous, and the conflicting group-level belief might *also* be autonomous. In this case, group members are alienated *as individuals*, not as group members.

We think this alienation threatens autonomy, but the threat is not to the autonomy of group members' *beliefs*. Rather, it threatens their autonomy as group members without undermining their epistemic autonomy. Nor does it undermine the epistemic autonomy of the group-level belief. Individual jury members in this sort of case believe that *p* (that the defendant is guilty) and do so autonomously. *Qua* jury, however, they believe not-*p*, and they may hold this belief autonomously too (after all, they may come to their verdict via processes with which they identify). As individuals, they may nevertheless feel alienated from the group belief.

We think much the same sort of thing is true of the heteronomy of the person who dissents from *p* but belongs to a group that believes *p* because a significant number of its members believe that *p* (as in the in the sort of case we sketched at the end of the last section). Indeed, this is a clearer case in which the alienation does not entail the heteronomy of the individual-level belief. After all, the individual who is able to sustain the belief that *p* in the face of group pressure is a paradigm of autonomy in belief. If that's right, then being alienated from the group's belief cannot threaten the autonomy of the individual's belief. Again, there's no belief from which the agent whose belief it is is alienated from that belief. If this sort of alienation threatens autonomy, as we think it does, it is not a threat to doxastic autonomy. It is moral autonomy that is threatened, although it is threatened by a doxastic route.

3.5.3 Distributed Accounts

We turn, finally, to distributed accounts. As we've seen, on these accounts, group members need not share their group's belief. For example, a scientific research team may defend the view that *p*, despite the failure of individuals to share that belief. It's easy to see how individual members might be alienated from, and unable to endorse, group-level beliefs in such cases. Consider a case in which a lab publishes a paper defending the view that masks do not prevent transmission of COVID-19. Some group members may be passionately opposed to that conclusion and therefore feel alienated from it. They may remain members of the group in good standing nevertheless, especially when they believe that the evidence was generated and processed appropriately.

In such a case, individuals may autonomously believe that *p* while belonging to a group that autonomously believes that not-*p*. There's no threat to epistemic autonomy at any level. Rather, individuals' alienation from the group-level belief might threaten their moral autonomy.

It's also easy to see that our distributed account allows for the possibility of a dissociation between individual-level belief autonomy and group-level belief autonomy. Since the group belief that *p* arises independently of individual beliefs about *p*, it does not impose any requirements on the latter. A group that (let's say autonomously) believes that *p* might be constituted of members who heteronomously believe that *p* (perhaps they've been oppressively socialized into the belief and yet produce good evidence for it, *qua* group, via reliable processes). In cases of this sort, the individual-level heteronomous belief may either match or conflict with the group-level belief.

The distributed account therefore allows for a dissociation between heteronomous individual belief and autonomous group belief. Can it allow for the reverse dissociation? We think it can. Although the groups we envisage adopt procedures for belief formation that are aimed at truth, these processes are no less open to manipulation by others than are the belief-formation processes of individuals. On the contrary, it is surely easier to, say, hack the data-processing routines of scientific groups than it is to engage in the sort of direct manipulation of neural processes that often feature in the thought experiments designed to isolate the properties of autonomous agents or beliefs. If an agent (individual or collective) interferes with such processes, the resulting group belief is manipulated, and its autonomy is undermined. Perhaps groupthink might be understood along these lines.[20]

We might also see autonomy threatened in cases where the belief-formation procedures are actually reliable. Suppose colonizers impose evidence-sensitive processes on the members of colonized group. The resulting belief might be heteronomous. One way to cash out the threat to autonomy here is via history-sensitive accounts of autonomy, according to which autonomous beliefs must be generated in a way that lacks certain properties, such as compulsion (Carter 2022).[21]

Might we see heteronomous group belief due to the failure to satisfy some kind of Frankfurtian identification condition? We confess to a moderate skepticism, at least with regard to the kinds of groups we actually see in contemporary societies. We don't think actually existing groups are, at present, sufficiently rich agents to be able to possess higher-order attitudes toward their own states. Perhaps a tighter integration might occur through new technologies (we can imagine a virtual network that constitutes a group agent in a rich enough sense). Such a group might be capable of the sort of alienation required for Frankfurtian heteronomy.

3.6 Conclusion

If we're right, group beliefs are as apt for assessment as autonomous or heteronomous as individual beliefs. Moreover, such assessment must proceed independently of the assessment of the autonomy of individual beliefs. Even *if* summativism is true about belief, such that a group cannot have a belief that diverges from the belief of the majority of its members, we may see double dissociations between the autonomy of the resulting beliefs.

Obviously, this sketch of the options is very schematic. Moreover, we make absolutely no claims to comprehensiveness. Quite the contrary: We strongly suspect there are other routes to group belief heteronomy, and other ways in which individual- and group-level autonomy might dissociate. Our aim has been to begin mapping the options. There's a lot more work to be done.

Not only do we need to do much more work to clarify the degree, and the mechanisms, whereby the autonomy of group beliefs might be threatened independently of that of individual beliefs, we have left untouched consideration of the normative upshots of our claims. For example, might there be a case for paternalistic intervention into groups that possess their beliefs heteronomously? Might groups be assessed for how well they've responded to their evidence, independently of how their members respond to evidence (it's worth recalling how utterly unimpressive, morally and epistemically, so many individual scientists are; nevertheless, they play their part in our most successful epistemic enterprises). We won't attempt to answer these sorts of questions here. We hope what we have done is enough, though, to motivate rescuing group doxastic autonomy from its utter neglect.

4 Epistemic Autonomy in the Age of AI

In this chapter, we aim to explore one context in which non-agential sources of testimony and guidance might threaten epistemic autonomy. Our focus will be on large language models, of the sort that have become familiar with ChatGPT, but we will also note how our argument might extend beyond LLMs.

Given the pace of development of LLMs, it's possible that they will quite soon themselves be agents. In one scenario, they come to be collaborators, and they raise no special issues with regard to epistemic autonomy—no issues, that is, that aren't already familiar from collaboration with other people. In another scenario, they will become vastly superior to us as knowers, and their superiority will raise a set of issues different from those we will discuss here. Indeed, some mainstream AI researchers think it is likely that the day when we lack the capacity to contribute meaningfully to cognitive activities in which they're engaged is *very* near (Kokotajlo et al. 2025). In either of these scenarios, what we say here is redundant (on the second, you would probably be better off asking the latest iteration of an LLM than reading this chapter). We assume that neither scenario is imminent. We by no means rule out the possibility of so-called AGI (artificial general intelligence) or even ASI (artificial superintelligence). We think it's probable that working with LLMs will not be like working with the human agents with which we're familiar: Either they will be cognitive tools, and not agents, or they will be agents with strengths and weaknesses unlike those we see in humans. For the most part, we focus on the issues that present-day LLMs raise, without much speculation on what the future holds.

We will also use the example of LLMs, and our interaction with them, to better understand the ways in which interaction with other human agents enhances, or threatens, epistemic autonomy.

4.1 LLMs as Scaffolding

For epistemically social animals like us, knowledge production is scaffolded in multiple ways. It is *socially* scaffolded: Knowledge production is distributed across agents and across time, such that we are capable of producing and transmitting knowledge that would otherwise be entirely inaccessible. Such social scaffolding is essential to the progress of science. Scientific knowledge is almost always produced by groups, not individuals. Collaboration is necessary, in view of the complexity of contemporary science. An individual scientist simply cannot know enough to avoid deep collaboration; a lab often consists of individuals with different sets of non-overlapping skills and expertise (one person might be an expert in statistics, another in the theory under test, a third an expert in the tools used, and so on). Increasingly, science is spread out *across* labs, ensuring that a diversity of perspectives and expertise are brought to bear. Peer review deepens the collaboration (even if it doesn't *feel* very collaborative): It ensures that the hypothesis and results come under hostile scrutiny, taking advantage of the way in which such scrutiny enables the identification of problems that those who are not motivated to find fault may miss. After publication, scientific findings come under even wider scrutiny. All of this is essential for the progress of knowledge.

Elijah Millgram (2015) has suggested that this deep interdependence, with its attendant risks,[1] is characteristic of modernity. Although it's plausible that science has deepened it, we think that deep interdependence is, in fact, characteristic of human knowledge. An epistemic division of labor is essential to human flourishing across the world and always has been. Although there may be relatively little division of labor *at a time* in traditional societies, there is a deep division *over* time. Traditional societies flourish because they use techniques and technologies bequeathed to them by many previous generations. These techniques and technologies often cannot be reinvented quickly or even within the lifespan of a single individual. Instead, they develop through proper evolutionary processes: through a ratchet effect (Tennie et al. 2009), whereby each generation builds on the achievement of the previous one—often without the capacity to reverse engineer the achievement. The resulting techniques may be quite opaque to those who use them: They may not know *why* the technique works or even what it achieves. Nevertheless, it may be central to their success (Richerson and Boyd 2008; Henrich 2015).

Knowledge production is also scaffolded in nonsocial ways. We use cognitive tools of various sorts to extend our epistemic agency. Our physical environment is structured epistemically. In medieval European towns, for example, church steeples typically towered over the skyline, making God salient. This might be seen as a kind of outsourcing of memory to the environment. Townspeople need not remind themselves of the importance of God because the environment bore the load for them. Then and now, roads do not merely make travel easier, they also serve an epistemic function. We need not recall the route to Larissa because we can simply follow the carefully delineated path (compare Haugeland [1998, 234] on *collaboration* with roads).

Present-day LLMs, which, since the release of ChatGPT-3 in 2020, have played an increasingly important role in knowledge production, might be seen as offering a sort of indirect social scaffolding of cognition. LLMs are trained on a very large corpus of text and, on the basis of patterns in the text (modified, to some degree, by pre-training and post-training), respond to queries by predicting the correct response.[2] Since the corpus of text is the product of countless individuals and groups, the output of a LLM is, to some degree, socially scaffolded. In interacting with a LLM, however, the training data fades into the background. We may know that the responses an LLM generates reflect its training corpus, and therefore the cognitive labor of many thousands of people over centuries, but we experience ourselves as interacting with a semi-agential artifact.

LLMs are already impressive, and they continue to improve very rapidly. They are already widely used across a variety of fields. For example, AI is already used in medicine for purposes ranging from the relatively banal, such as scheduling (Knight et al. 2023), to the interpretation of radiographic images (Palm et al. 2022; Eisemann et al. 2025), and daily clinical practice (Singareddy et al. 2023). They are transforming professions as disparate as journalism (Merrefield 2025), school teaching (Shroff 2025), and the law (Richter 2025). How might the many and increasingly diverse uses of LLMs affect agents' epistemic autonomy?

4.2 Disagreeing with Others, Human and Otherwise

At their current level of development, LLMs are best thought of as cognitive tools. As such, they promise to enhance epistemic agency. Increasingly, for example, they allow us to utilize much larger datasets, to identify patterns that we would not otherwise have seen, to bring different sources of information into dialogue with one another, and to make new discoveries. Although this represents an enhancement of our epistemic powers, this enhancement need not entail, or correlate with, an increase in epistemic autonomy, however. More powerful agents are not necessarily more autonomous agents.

Nevertheless, epistemic power might well *conduce* to epistemic autonomy. A more powerful epistemic agent might be better able to resist manipulation, for example. We might utilize LLMs to check the sources of information in real time, flagging those that have been identified as unreliable in the past. If there are fingerprints of misinformation, as some claim (Van der Linden 2023; see D. Williams 2023 for dissent), an LLM might be able to quickly identify them. More powerful LLMs will be able to place information in a broader context, making it harder for others to deceive us with true but misleading claims.

LLMs might also enhance epistemic autonomy in more unexpected ways. Sometimes, interaction with an LLM might be autonomy enhancing compared with interaction with an equally capable human agent, independent of any concerns about manipulation. Consider, in this context, peer disagreement as it usually occurs with "disagreement" with an LLM at its current level of capability.

In familiar cases of peer disagreement, two agents who are equally well equipped to judge a certain matter come to divergent assessments of it. The recognition that an epistemic peer disagrees with me whether *p* represents epistemic pressure to conciliate: to reduce my confidence in my judgment. Assuming the question is truth evaluable, either I or my peer (or both, of course) has made some kind of error. Since they are my peer, I have no reason to think it is more likely that they have erred than I have. These considerations motivate the equal-weight view: Ceteris paribus, when agents find themselves in such cases, they ought to give equal weight to their own judgment and that of their peer and—accordingly—suspend judgment whether *p* (Christensen 2007; Elga 2007).

Marušić and White (2023) have recently suggested that the equal-weight view entails alienation from one's own beliefs. They argue that to confer equal weight on my assessment of the evidence and my epistemic peer's conflicting assessment requires me to regard my belief as a mere psychological fact about me. But I can see my own judgment as a mere psychological fact about me only by objectifying my own mental states, and that's a profoundly alienated view of one's own belief. Reasoners, people who take the participant attitude (Strawson 1962) to one another, cannot treat their own beliefs (or indeed the beliefs of others) as mere psychological facts. Rather, reasoners set against the beliefs of their peers not their final judgments, but the very considerations that constitute their evidence for these judgments (together, presumably, with the inferential relations one detects between these considerations). To reify our belief in the way the equal-weight view requires is to see it as an arational extrusion, not the endpoint of the ineliminably agential activity that is reasoning.

As an alternative to this alienating perspective on peer disagreement, Marušić and White (2023) suggest mutually answerable shared inquiry as the right response in such cases. We, too, value such shared inquiry. We agree that peer disagreement often calls for reasoning together.[3] In the restaurant case, for example (Christensen 2007), it is, of course, reasonable for you and me to recheck the sums, perhaps swapping our work, to see where at least one of us went wrong. But proponents of the equal-weight view certainly need not denigrate such shared inquiry, nor do they deny that checking is a good response to peer disagreement. Rather, they maintain that *prior* to such checking, and when it is impossible or has failed to resolve the disagreement, we should conciliate.

We reject the claim that unalienated agents set the considerations that constitute their evidence, rather than their final judgments, against the judgments of their peers. In typical cases, agents simply cannot do so. We can't do that because we aren't able to encompass the considerations that have served as our evidence, let alone the inferential relations between them, in working memory. We are unable to have a Cartesian "clear and distinct grasp" of these considerations. Consider the restaurant case again: At the moment I discover that my peer disagrees with my assessment of what each of us owes, I can't recall what every entree cost, let alone the subtotals. Further, I recognize that the sum is difficult enough to perform in one's head for mistakes to be reasonably common. Since I can't encompass my reasons in working memory, I can't set them against my peer's judgments, or their reasons. I must see my judgment as defeasible, and my peer's judgment as evidence against it. There's nothing alienating about that, we maintain.

There *are* cases of (apparent) peer disagreement in which the equal-weight view would require us to adopt an alienated stance toward our own judgments. In cases in which we can encompass the considerations that form the basis for our judgments in working memory, seeing our judgment as mere seemings would require us to abandon the participant perspective. I can't see "two plus two equals four" as a mere cognitive seeming. But as Jennifer Lackey (2010) has pointed out, in cases such as these, there's no rational pressure to conciliate. Instead, we relegate our interlocutor from the status of epistemic peer.[4]

All that said, once we have shared all our evidence and checked all our reasoning, as well as our peer's arguments for their conclusions and *their* evidence, if we continue to disagree, we find ourselves in an epistemically uncomfortable position. Even if it is appropriate to see peer disagreement as higher-order evidence against our judgment that *p*, we can't take the same attitude to such higher-order evidence as evidence that directly bears on whether *p*. The rational pressure we are under to treat our peer's judgment as a strong piece of evidence against our own judgment does induce something like a feeling of alienation. We think, however, Marušić and White (2023) have misdiagnosed the source of unease, and that contrasting peer disagreement with "disagreement" with an LLM—at least at their current level of development—allows us to better identify the source of the unease.

Suppose the total you reach at the end of a series of sums conflicts with the total that an LLM reaches. This case is, of course, identical to the restaurant case, with the exception that the agent is no longer disagreeing with a person. If Marušić and White's (2023) diagnosis of such cases were correct, you would experience alienation from your judgment if you saw the LLM's output as evidence against it. We think it's obvious that you would not hesitate in seeing its output as strong evidence (perhaps even dispositive evidence) against your judgment. But we also think it's clear that you would *not* experience a sense of alienation in doing so. This case therefore suggests that it's not the attitude that peer disagreement forces us to take to our own mental states that leads to something akin to alienation. Rather, it's because peer disagreement is *interpersonal* that it gives rise to unease.

We think "alienation" may not be quite the right term for the unease experienced. Rather, peer disagreement places in question our standing as epistemic equals. When I disagree with a peer, I stake something not only on the upshot of my reasoning but also on its quality. Although different people may value reasoning to varying extents, we think it's at least common, and perhaps universal, to identify oneself, at least in part, with one's reasoning—that is, to see oneself as an essentially rational being. To the extent to which peer disagreement throws our status as reasoners into question, it is disturbing. The more basic the error that seems to be at issue, the more our status as rational agents is in question.

Peer disagreement is a threat to our standing as rational agents because it is disagreement with *peers*: people whom we see as our equals. When we value our capacity to reason well about a particular topic or in a particular domain, having our reasoning shown to be inferior to theirs threatens our relative status. Conversely, when we find ourselves disagreeing with someone who is, on this particular issue, an acknowledged epistemic superior or when we don't care very much about our capacity to reason well in a particular domain, we experience little to no unease in treating the conflict as evidence that we've gone wrong in our reasoning.[5]

Although we don't think Marušić and White (2023) are quite correct in their labeling of the feeling, they are quite right in pointing to some sort of unease often experienced by agents in cases of peer disagreement. We want to suggest that this experience may, in some circumstances, represent an obstacle to the achievement of epistemic autonomy.

Autonomous belief is an achievement. It requires the development of capacities—the capacity to assess evidence carefully, for example—that require an investment of time and effort. Token autonomous beliefs are themselves often achievements, requiring the application of these capacities anew. Autonomous belief requires a degree of self-trust: The agent must trust that she's capable of answering the question for herself. In the absence of such self-trust, she will lack the confidence to come to her own judgments or to maintain them over time.

The capacity to cope with peer disagreement, to conciliate when it's called for and to engage in deliberation with others, is essential to the development of the capacity for autonomous belief. An agent who folds too quickly in the face of disagreement is by no means autonomous. On the other hand, the agent who is stubborn in her beliefs passes up opportunities to learn from others that are not merely important for the justification of the belief at issue but also (perhaps more importantly) essential for the development of the capacities she needs to be an autonomous epistemic agent. That being the case, it would be quixotic to call for agents to avoid peer disagreement. But peer disagreement, at the wrong time and in the wrong circumstances, can be an obstacle to the development of the self-trust needed for autonomy.

Peer disagreement is no particular threat to novice learners: They should not and typically do not feel threatened by dissent, since they don't stake their self-trust on their capacity to evaluate evidence and reason well regarding the matter at hand. But for more advanced learners, peer disagreement, especially repeated peer disagreement, may begin to generate unease. They may worry that despite their hard work, they are unable to take the next step—that they lack a talent or insight their peers may have, or that they lack the discipline to assess the evidence well. In this kind of context, we suggest, the use of LLMs might be a means for better learning and the inculcation of the self-trust required for epistemic autonomy.

Peer disagreement may interfere with the development of the self-trust required for epistemic autonomy. But learning to handle peer disagreement well is *itself* a necessary component of the capacity for autonomy. Epistemic autonomy requires the capacity to think for oneself—and is therefore incompatible with the spinelessness of the person too ready to conciliate—without, however, slipping into stubbornness. We think that LLMs, so long as they remain cognitive tools and not agents, might help with the development of the requisite capacities because disagreement with this sort of cognitive tool is experienced as less threatening than disagreement with an agent.[6] They enable practice in a relatively unthreatening environment. An LLM calibrated to the skill level of the learner will not only serve as a check on her results but also enable her to develop the skills needed for coping with disagreement. LLMs provide a safe environment in which to take epistemic risks and thereby, in effect, practice being wrong.

Of course, the learner also needs to confront the challenge of disagreement with a genuine peer; the LLM cannot do away with the need for interaction with human agents. But LLM disagreement might scaffold the acquisition of the skill of accepting that we are wrong.

We suspect, but won't argue, that the point can be generalized: LLMs may be helpful to the development of the capacities required for epistemic autonomy more generally because they enable learning in ways that are both tailored to the skill level of the learner and unthreatening to her growing self-trust.[7]

Whether LLMs are, on balance, a net plus for epistemic autonomy remains to be seen, however. They may also undermine such autonomy in other ways.

4.3 Agency Lost?

LLMs play a variety of epistemic roles—producing testimony (more or less reliable), transforming data, generating and evaluating ideas, and so on.[8] In many ways, interacting with LLMs is already like collaboration. One person who uses an AI in designing components for spacecraft describes working with it as "like collaborating with an alien" (Paris and Buchanan 2023; compare Sourati and Evans 2023) to capture its peculiar combination of unexpected insights and inexplicable blunders. In the previous section, we suggested that some of the ways in which collaboration with an LLM is different from collaboration with human agents may help to protect autonomy. In this section, we explore ways in which such collaboration might undermine epistemic autonomy.

Rini (2023) suggests one way in which such a threat might arise. On her Kantian conception, the self-legislation required to be an autonomous agent essentially involves the recognition that other agents are equally self-legislating. The practical point of view, the point of view that we must adopt to see ourselves as autonomous agents, can be adopted only from within a framework that recognizes the autonomy of others. From this Kantian perspective, Rini sees a threat from interaction with LLMs.

The threat arises because regular interaction with LLMs can be expected to inculcate a specific set of dispositions. LLMs already feel quite agential, and the extent to which interaction with them feels like—or, indeed, becomes—interaction with an agent will likely rapidly increase. But barring the kind of singularity that so obsesses many writers on AI, they will remain subservient to us.[9] When we interact with them, we will experience ourselves as acting with an agent with whom there is no reciprocity—whom we wholly dominate.

Some people describe their interactions with LLMs as less like "collaborating with an alien" and more like "collaborating with a slot machine" (Michaels 2023).[10] So long as it feels like *that*, Rini's worry won't arise. It's when and because the exchange feels agential that it risks inculcating the dispositions she's concerned about. LLMs are "pliable agential simulacra," in her terminology. Because they're pliable agential simulacra, they're "autonomy traps": They tempt us into exercising a petty dictatorship. It's a dictatorship because the LLM does, or at any rate tries to do, exactly what we ask of it. It's an autonomy trap because interacting with an LLM feels like interacting with an agent, and therefore exercising dictatorial rule over LLMs tends to inculcate nonautonomous dispositions. If LLMs become genuine agents in the future, but remain subservient (or even sycophantic) nevertheless, the threat will increase.

So long as LLMs are not themselves agents, the degree to which the danger Rini points to is genuine depends on an empirical question: Are the dispositions inculcated through interacting with LLMs generalized to interactions with agents? Rini assumes that the fact that the first such kind of interactions *feels* agential ensures such generalization, but that may well not be the case. The knowledge that LLMs are not agents may block generalization. Of course, LLMs may come to be genuine agents. If they do, and they remain subservient at this point, then it may be wrong to exercise petty dictatorship over them, although the dispositions *still* may not generalize to human agents (we suspect that the danger Rini points to arises more seriously for run-of-the-mill contemporary employers, who may exercise a petty dictatorship over people they know full well to be autonomous agents, rather than for those who interact with LLMs).

We don't share Rini's Kantian account of autonomy. Although the danger she points to may be real—it may be corrosive of moral character to exercise a completely one-sided authority over something that feels agential—we think that the danger this represents to autonomy, and specifically to epistemic autonomy, is relatively peripheral. We are more concerned with a very different danger for epistemic autonomy arising from our interactions with LLMs.

Rini envisages the problem as arising from the way in which people *treat* LLMs. Our concern focuses on the actual status of LLMs. Now, and in many future scenarios, their status is unclear. Are they cognitive tools—"slot machines"—or are they genuine collaborators in our epistemic enterprises? Neither option is itself worrisome. If LLMs are mere cognitive tools, then we should expect them to be incorporated into our epistemic agency in much the same way in which such tools are routinely integrated. That's not to suggest that LLMs won't offer novel opportunities to us—they surely will (they already do). We don't yet know with any clarity which of the tasks we currently perform LLMs may largely or entirely do for us. We have even less of an idea what new cognitive horizons LLMs will open up for us. But these open questions concern the *content* of the new knowledge we will generate with them and the *scope* of the agency we will exercise using them, not how they will affect epistemic autonomy. The integration of cognitive tools into our epistemic lives promises the extension of our agency and doesn't threaten autonomy.

Similarly, if LLMs are or become genuine collaborators, we will interact with them in familiar kinds of ways. If LLMs become our collaborators, the possibilities they open up might be even more exciting and even harder to predict. But collaboration is the most familiar thing in the world; what matter, as regards epistemic autonomy, that we collaborate with an LLM?

But LLMs today, and in the future on many scenarios, do not fit easily into either the tool or the collaborator model.

The two quotations we used from people who are already working with LLMs illustrate the ambiguity they present us with. Is working with them more like "collaborating with an alien" or like "collaborating with a slot machine"? Are LLMs genuine independent contributors to our epistemic projects, perhaps even helping to define the goals at which we aim, or are they mere tools? Rini's worry arises only if we experience interaction with LLMs as interaction with an agent. But their very subservience, as well as their limitations and oddities, may prevent us from experiencing this interaction as truly agential. On the other hand, however, perhaps their (apparent) intelligence may prevent us from engaging with them as mere tools. This ambiguity might *itself* represent an obstacle to the epistemically autonomous use of LLMs.[11]

We can illuminate the potential danger here by focusing on the credit due to agents for their cognitive achievements. The appropriateness of credit is often used by epistemologists as a guide in assessing whether a belief amounts to knowledge or whether its formation manifests epistemic virtue. In making this assessment, epistemologists ask *who really deserves credit for a putative epistemic achievement*? We want to focus on a slightly different question: *Who does the agent think deserves credit for the putative epistemic achievement*? In asking this question, we focus on one component of epistemic autonomy: the ownership condition.

On many views, for an agent appropriately to be *held* responsible for a belief, they must have taken responsibility for it—that is, they must satisfy an ownership condition. The ownership condition is often motivated by science fictional thought experiments: desires implanted in agents by brain washing, gene editing or selection, or direct intervention in the brain, such as Al Mele's Ann/Beth case (see Mele [2013] for a presentation of the case and discussion of some of the literature that has grown up around it). Here's a short version of the case:

> Ann and Beth are employed by the same philosophy department. Despite their more or less equal talent, Ann is much more productive than Beth. The difference arises from the fact that Ann (unlike Beth) is dedicated to philosophy above all else; Beth enjoys philosophy, but she enjoys many other things just as much. Their dean wants Beth to be as productive as Ann. To this end, he employs heavy-duty brainwashers to install Ann's hierarchy of values in Beth while she sleeps and without her knowledge. Beth is surprised to find herself valuing philosophy above all else on awakening, but she endorses the change, surmising she has matured into it, and dedicates herself to the pursuit of philosophy.

Mele and many others who have discussed this sort of case argue that if central desires, motivations, values, or ends are implanted in us without our consent, we are not autonomous with regard to them, and actions that owe their character to such states are not autonomous actions.

Mele is careful to point out that the fact that a desire, value, and so forth is implanted does not entail that the agent can never autonomously perform an action that owes its character to that state. Agents can come to take responsibility for their values and commitments, however they came to have them in the first place. That had better be the case. After all, we each acquired many of our beliefs and our values from our parents and surrounding society when we were children and prior to our having the ability to appreciate them and to reflect on them. Autonomy develops over time. Young children have little autonomy (and correlatively cannot be held responsible in any robust way), but in the normal course of events, they come to be autonomous adults. We come to be autonomous, in the main, by living with our values and ends, responding to feedback from the world and from other agents. Such feedback gives us the opportunity to reflect on our acquired commitments and thereby to work on them. We modify some values, reject others, and deepen our commitment to yet others. Over time, Beth (too) may come to endorse her new commitment to philosophy and hold it autonomously. Over time, she may instead reject it. In either case, she takes responsibility for it, and thereby comes to be autonomous with regard to actions and states that reflect it.

Agents may, therefore, fail to satisfy the ownership condition in cases in which their values or beliefs do not arise from their reasoning due to a lack of time for reflection. Beth may come to be autonomous with regard to her implanted values, but the very first actions that express them are not autonomous because she has not had time to make these values her own. But agents may also fail to satisfy the ownership condition, even given ample time for reflection, if their values or beliefs are strongly resistant to deliberative assessment. If valuing philosophy is in a literal sense hardwired into Beth's brain, such that no amount of dissatisfaction with it would weaken its hold on her, then she cannot ever be autonomous with regard to it. (Mele calls such values "unsheddable": Agents cannot be autonomous with regard to their unsheddable values.)

Of course, the heavy-duty brainwashing imagined by Mele does not exist, and beliefs and values are never literally hardwired into our brains. But there are plentiful actual cases in which agents come to acquire values and beliefs that are unsheddable and which therefore motivate an ownership condition on autonomy.

Agents may have values that are strongly resistant to reflection when they are the victims of *oppressive socialization* (Benson 1991). "Socialization" is the general term for the developmental acquisition of values from our society and culture in normal development. Socialization is oppressive when the values and commitments that are acquired in this way are packaged in such a way that they cannot be rejected.

There are several different ways in which socialization might result in values or goals that are, for practical purposes, unsheddable. Unsheddable values can be inculcated through the deliberate short-circuiting of the very reflection that could lead to the agent taking responsibility for them. For example, some religious fundamentalists oppressively socialize children into their version of their faith by linking any kind of questioning of its foundations to guilt or the fear of damnation. Oppressed or marginalized people might also be oppressively socialized by instilling into them the value of self-deprecation or obedience. It is difficult to question such values because such questioning requires putting them aside. More commonly, perhaps, people might be oppressively socialized into any values whatsoever simply by failing to equip them with the tools for effective reflection: Patriarchal societies steer women away from education, thereby making it more difficult for them to reflect on their values.

Values may also be instilled in a way that makes them unsheddable by taking advantage of the formation of adaptive preferences (Stoljar 2014). Preferences are adaptive when they enable the person to cope with the limitations their circumstances impose on them. Some such adaptation is appropriate—for example, the person who has the goal of becoming a professional basketball player might appropriately respond to the realization they lack the physical attributes required by preferring to develop other talents and goals (Nussbaum 2000). This sort of adaptation is a healthy response to circumstances for which no one is to blame. But people appear to adapt their preferences to oppression too. Doing so is in their best interests so long as changing their oppressive circumstances is not an option for them, but adaptive preference formation may lead to oppressed groups overlooking opportunities for such change. Plausibly, the agents are less than fully autonomous in holding these values, even when they are genuinely adaptive responses to the limitations of their circumstances.[12]

Agents who are the victims of oppression socialization or who come to their values through adaptive preference formation are not fully autonomous, and a major part of the reason for their lack of autonomy is that their values and commitments are not in the fullest sense their own. These values and commitments do not reflect agents' own rational assessment of their evidence; rather, they arise through processes that bypass deliberation and are resistant to reflection once inculcated.

We think that the unclear status of LLMs, now and in many future scenarios, give rise to unique issues with regard to the satisfaction of the ownership condition on epistemic autonomy.

We have suggested that LLMs are currently experienced as partially agential and partially tools. They are poised between slot machines and intelligent aliens. They are agentially ambiguous, we will say. This agential ambiguity affects many token instances of their deployment. When someone uses them to generate ideas (or lines of poetry, as Sean Michaels did), their experience of the interaction hovers between the experience of interacting with an independent source of agency and the experience of using a tool. Agential ambiguity might be greatest when we engage in dialogue with an LLM. Their responses to questions and to follow-up refinements and clarifications not only seem agential but also exhibit limitations and quirks that do not seem agential at all (consider their misplaced eagerness to apologize upon being told they're mistaken when the person interacting with them "corrects" them by repeating exactly the same information the LLM itself provided).

We suggest that agential ambiguity is an obstacle to satisfaction of the ownership condition on epistemic autonomy. Here, the threat arises not in the formation of values or preferences (although, of course, long-term interaction with LLMs might shape values or preferences) but rather with regard to *outputs*: to ideas generated, products designed, papers written, and so on. The (human) agent who interacts with an LLM cannot fully take responsibility for their own contribution to the project because agential ambiguity is an obstacle to identifying just what that contribution consists in.

There's no special difficulty in identifying our contribution to a project arising from the use of tools. Tools are extensions of our agency. The fluent use of tools produces outputs that are fully ours. The tool–person dyad should be considered a single intelligent agent. If LLMs were just tools, they'd present no special obstacle to satisfying the ownership condition.

Collaborators are not integrated into agency in this way, of course: When we collaborate with others, we do not form superorganisms. That's not to say that groups of collaborators might not themselves be agents. In the last chapter, we suggested that, under certain conditions, groups might well be considered epistemic agents. Here, though, we're concerned with the *individual* autonomy of contributors to the group. Although other organisms, such as the eusocial insects, might form superorganisms composed of individuals who do not count as agents in their own right, group agents—if there are any—are composed of individuals who are *also* independent agents. As individual agents, their autonomy can be threatened or undermined independently of any group agent they might constitute (and independently of whether the group agent counts as autonomous).

Collaboration with other agents also need not introduce any special threat to individual autonomy (although, as we will see, it might well introduce significant complications with regard to it). In a collaborative group that preserves individual epistemic autonomy, each agent understands their own contribution to the project. The boundary between what I do, and in particular my reasons-responsive contribution to the group, and what others do in such a group is relatively clear. But individual autonomy is threatened by agential ambiguity.

Agential ambiguity prevents us from identifying what is the product of our agency, albeit extended and enhanced, and what (if anything) is owed to the contribution of other agents as well or instead. For what do *I* deserve credit? Indeed, do I deserve credit at all? Perhaps I am the cognitive tool, in this context, and the LLM is the only individual agent at work (at least with regard this output). The inability to identify the extent to which the overall product reflects my agential contribution prevents me from taking responsibility for it: Since I do not know whether I deserve credit, I cannot see the product as the result of my reasons-responsive agency.[13]

In cases in which the LLM is a dispensable scaffold, the agent can evaluate the output once it is produced, deciding whether to endorse it and, to that extent, taking responsibility for it. But taking responsibility for the *product* does not entail being responsible for its *production*. We can take responsibility for the outcome of other agents' actions (e.g., when they work for us), but that does not entail we deserve credit for the process of production. In some cases, we may not even be able to take responsibility for the outcome. Sometimes, the output might be inscrutable to the agent without the assistance of the LLM. In those cases, the agent can take responsibility for neither the production nor the product (still less can they modify or reject it for epistemic reasons). In all these cases, the agent's epistemic autonomy is threatened.

4.4 Authorship Is Increasingly Amorphous

Although collaboration with agents need not threaten individual epistemic, there are cases in which much the same issues seem to arise without the need for LLM involvement. There are cases of collaboration with human agents in which we lose the capacity to identify our own agential contribution. Is our capacity to take responsibility for the production or for the product threatened in these sorts of cases?

One possible case involves work by long-standing collaborators with the collaborative project extending across significant periods of time. Such collaborations might begin with each person making distinct contributions, but over time, the collaborators may lose track of just who has done what.

Ideas might be generated in dialogue, with no clarity over which agent made what difference to the evolution of the idea. Sentences might be written and rewritten by several hands, so that the result is an ineliminably joint product. Collaborators typically do not experience this sort of work as a threat to their epistemic autonomy—on the contrary, they may feel especially proud of it.

We think these sorts of cases are crucially different from agential ambiguity cases. In agential ambiguity cases, we cannot discern whether the product and the production manifests our agency alone or joint agency—that is, we cannot discern whether the LLM is essentially a tool or whether it is, to some significant degree, another agent. In fully collaborative cases, we can rule out one of these options: We know that the other agents are not mere tools and that therefore the product does not manifest our agency alone. Any ambiguity comes at the other end of the scale. In large collaborative groups, and especially when the process is extended and involves many rounds of revision, we may wonder whether it reflects our contribution to any significant degree at all. In many disciplines, papers with dozens, even hundreds, of coauthors are common. A contributor to such a paper might wonder whether the final version bears their mark in any appreciable way.[14]

In some cases, the agent will be able to identify some contribution they have made to the product. In such cases, the agent knows that they stand in a fully agential relation to the product, even if they cannot identify the extent of their contribution to it. The knowledge that we stand in such a relation to it allows us to take responsibility for it. We can take full responsibility for it *as* a joint product. In such collaborations, the product is *ours*, not *mine* and *yours*, but it reflects and manifests our individual agency. In collaboration with an LLM, we cannot even be confident that the resulting work is *ours*: Whether another epistemic agent was involved at all is in question.

In some cases, though, deeply collaborative activity might reduce epistemic autonomy in a way that parallels the threat to epistemic autonomy that arises from work with an LLM. We may proudly take responsibility for the product, but when (as is increasingly common) it is the work of many people with different skills, our individual capacity to reflect on it is greatly reduced. Insofar as the product is genuinely ours, and reflects our joint agency, the individual's capacity to endorse it, to alter it, or to reject it may be reduced. In actual cases, the agent is often unable fully to appreciate the intellectual underpinnings of the project, since they lack the full set of skills they'd need to grasp the full justification of every inference and every piece of data (sometimes they cannot grasp the justification of some move at all). Even when the agent is confident they have made a fully agential contribution to the final product, their relation to it may lie somewhere between the endorsement of someone who grasps a claim for herself and that of the person who endorses it on the basis of testimony.

The degree of reduction in epistemic autonomy that arises in this sort of deeply collaborative work is a function of the agent's capacity to grasp the whole. In contemporary scientific collaboration, where work is often distributed across many labs and reflects the contributions of many different disciplines, inability to grasp the whole may be the norm. Different group members collaborate using methods of which others within the group have little grasp and little capacity to verify for themselves.

In this sort of case, an individual researcher is typically justified in believing the conclusion the group reaches, but her grounds for justification are not dissimilar to the grounds for accepting the claim that a well-informed reader of their publications might possess. Her intellectual activity is responsible for only a small part of the overall product. Even in those cases in which it is essential to the justification of the claim, it will almost invariably fall very far short of establishing it. She accepts the rest on the same sort of basis that an outsider might: because it is asserted by people she has reason to trust and because she has good evidence that it was produced through reliable methods. She autonomously accepts the conclusion the group reaches, but her capacity to reflect on it and thereby take responsibility for it is relatively limited. This in no way reflects *her* limitations as an individual. She might be as well positioned as anyone could be to assert the conclusion her research group comes to, and she might possess as good a grasp of the overall evidence as any individual could have. Rather, it is an ineliminable reflection of the deeply distributed nature of the research.

In this sort of environment, it may be no exaggeration to say that there is no individual who stands in the right sort of relation to the research to take responsibility for it—no one who can vouch for it (Kukla 2012; Winsberg et al. 2014). That is, no one stands in the traditional authorship relationship to its claims; rather, they stand in much the same relation to it as do well-informed outsiders. Correlatively, no one may deserve epistemic credit *for the conclusion of the group*. Each person deserves credit for their contribution to the work, but that contribution is insufficient for the conclusion. (Indeed, there are many cases in which agents who appropriately are *not* credited as authors of a paper nevertheless made a bigger contribution to the work than many who are: when the former developed intellectual apparatus that the paper uses, for example.) That's the current state of play as we see it. As LLMs are increasingly utilized within science, the resulting agential ambiguity may further threaten the ability of agents to take responsibility for the work they help to produce.

4.5 Tools, Collaborators, or Some Secret Third Thing?

If agents often do not stand in the right sort of relation to some of the most important and impressive intellectual achievements of humanity to count as intellectually autonomous with regard to them, perhaps we should question the value of epistemic autonomy. Perhaps an inability to take responsibility for joint productions of LLMs and agents might not be especially worrisome. What matters, we might think, is the value of the output rather than in what relation to it the producers stand.

We think there's something to this line of thought, but it needs to be unpacked carefully before we're willing to endorse it. There are good grounds to think that the relation in which people stand to beliefs and belief formation that we honor by describing them as epistemically autonomous is a valuable one. But it does not follow by any means that agents should strive to maximize the number of beliefs with regard to which they stand in that relation, nor that there is anything epistemically suspect or second rate about not being able to stand in that relation to much of our knowledge. We ought to want to play an epistemically autonomous role in the production of knowledge, but we need not fall short if we do not stand in that relation to the product itself.[15] If we stand in the relation characteristic of the informed outsider, that may be autonomy enough.

We might one day, perhaps soon, find ourselves collaborating with agential LLMs. If and when we do, we will find ourselves in the same relation to them as we now stand to our human collaborators. They will contribute to our projects in ways that we cannot fully grasp for ourselves, resulting in products for which we cannot take overall responsibility. But that's not a new threat to, or limitation of, our epistemic autonomy; it's merely an extension of our existing relation to deeply collaborative work.

If LLMs are mere cognitive tools, they represent no obstacle to epistemic autonomy at all. If we're right, though, LLMs are not (yet) full collaborators, nor are they mere tools. They're agentially ambiguous, and agentially ambiguous LLMs do represent an obstacle to epistemic autonomy. If we are unable reliably to discern to what degree our agency is at work in the production of knowledge, we cannot take the responsibility required for our contribution to be epistemically autonomous. LLMs might disrupt epistemic autonomy in a way that deeply distributed knowledge production does not. It remains an open question, we think, whether or to what degree that's, on balance, a bad thing.

5 Negative Autonomy, Dependence, and Epistemic Axiology

5.1 Introduction

This chapter picks up on a theme canvassed in chapter 1—roughly, the relationship between epistemic autonomy and epistemic value and how epistemic dependence relates to both. Recall again our initial cautioning against the Marlboro Man model of autonomy. This, we suggested, was not any kind of ideal. The most extreme form of the epistemic Marlboro Man is a caricature of Emerson's (1841) hero in *Self-Reliance*: Such a person will—rather than consult a smartwatch—attempt to fashion a sundial when they need to know what time it is. And they will not accept as true claims about the existence of any city they've not yet visited and seen, presumably just whatever cities they find by developing their own mapping system and so forth.

Such a caricature goes in, as it were, "whole hog" when it comes to fetishizing specifically *negative epistemic autonomy*—namely, intellectual *independence* from external influence or aid in the way one's beliefs are formed. This kind of caricature lines up with an approach to inquiry that Thomas Reid warned us about. Such a person simply won't know very much at all,[1] as they'll be, at every opportunity, sabotaging themselves epistemically, as a matter of principle, just for the sake of epistemic independence from others' cognitive labor.

Let us set this cartoonish version aside and consider a more cautious and potentially philosophically interesting articulation of this ideal. This "enlightened" epistemic Marlboro Man acknowledges Reid's point that aiming for absolute independence in inquiry is a recipe for epistemic disaster. Their commitment to autonomy is therefore qualified by a critical ceteris paribus clause. This more cautious thinker aligns with the following pair of evaluative and prescriptive norms, which are designed to, in short, control for Reid's worry:

Negative epistemic autonomy thesis (evaluative norm): $\text{NEA}_{\text{evaluative}}$: Ceteris paribus, for any subject S and question Q (whether *p*), S's answer to Q is more *epistemically valuable* to the extent that it is arrived at through *intellectually independent inquiry*.

Negative epistemic autonomy thesis (prescriptive norm): $\text{NEA}_{\text{prescriptive}}$: Ceteris paribus, for any subject S and question Q (whether *p*), S *ought to prioritize intellectually independent inquiry* when answering Q.

These qualified norms don't praise and prescribe a policy of solving our intellectual problems under epistemic quarantine come what may. The force of each is entirely constrained by the "all else equal" caveat. In many situations, particularly those where what we're after is specialized knowledge of the sciences, all else is obviously *not* equal, in the following sense. Consider a citizen weighing the safety of a vaccine. Suppose that while if they defer to Dr. Fauci, they'd have a reliable path to a true belief, "doing their own research" with a search engine is likely to lead to a false one; here, the ceteris paribus clause is simply not met. In such a scenario, deference is really the only knowledge-conducive path, and so $\text{NEA}_{\text{prescriptive}}$ is silent. Likewise, $\text{NEA}_{\text{evaluative}}$ assigns no "epistemic value demerit" to the deferential belief, as there is, let's assume here, no more valuable, independent route to the same knowledge with which to compare it. With this in mind, consider that one could, in principle, outsource the vast majority of one's beliefs to experts without actually violating these norms, provided that doing so is the only viable path for them to knowledge.

That said, the fact that these norms simply lack application in cases where independent inquiry but not deference or outsourcing would lead to error does not itself make them trivial. As inquirers in information-dense environments, we often face a choice between multiple zetetic strategies we might employ for answering a given question we have—namely, strategies that vary, potentially significantly, in the degree of epistemic dependence they would require. Consider some mundane examples. Suppose you want to know your heart rate. To find that out, you could manually take your pulse (e.g., press on your wrist or neck with your fingertips until you feel the pulse, then count the beats for fifteen seconds and multiply by four), or you could just trust your smartwatch. Both methods work. To ascertain the distance to a water hazard on a golf course, you could pace it out by physically stepping the full distance, or you could use a digital rangefinder. To know if the bank is open, you could check its website or walk over to the building and look through the window to see if people are actually being served.[2] In all these cases, multiple paths to the truth are available, some more independent than others. It is in precisely these common scenarios that the NEA theses gain their purchase. The million-dollar question, then, is whether there is any truth to these norms.

Both $\text{NEA}_{\text{evaluative}}$ and $\text{NEA}_{\text{prescriptive}}$ have found some sympathy in contemporary epistemology. The most compelling, although often implicit, support for them arises from arguments that contrast solitary, independent inquiry with simple deference. This chapter will focus on two prominent examples of this strategy in the work of Ernest Sosa (2021) and Duncan Pritchard (2016). Although we agree with Sosa and Pritchard when it comes to acknowledging the distinct value of *understanding-why* (something not ensured when one simply knows by sheer deference or through extensive outsourcing),[3] we will argue that tying this value too tightly to the value of first-order epistemic independence—namely, of negative epistemic autonomy—is a mistake. We will advance two lines of argument to demonstrate this. First, we challenge the extensional adequacy of the NEA theses by examining cases of "mixed-methods" inquiry, which reveal that valuable epistemic achievements are often secured through deeply dependent methods.[4] Second, we press an overgeneralization objection against $\text{NEA}_{\text{prescriptive}}$ by arguing that it leads to a reductio ad absurdum that exposes a problem in its underlying axiology.

These arguments, we contend, show that the very idea (intuitive as it might seem initially) of epistemic value robustly tracking (as opposed to just accidentally correlating with) negative autonomy in one's belief formation is a dead end.

5.2 A Case Pair

To bring the prima facie appeal of the negative autonomy theses into sharp relief, consider a classic case pair that pits solitary inquiry against testimonial deference:

Case A: Mathematician A works alone in her study, proving a theorem, T, using only her biological cognitive faculties. She consults no one and uses no external aids.[5] In doing so, she comes to know that T is true.

Case B: Mathematician B can't be bothered with all the faff of working through the problem, so she takes a shortcut. She looks up the answer in the back of an authoritative textbook and, trusting the author, forms the belief that T is true.

Both mathematicians achieve exactly the kind of intellectual success they sought. Even more, they both acquire (at least) knowledge that T is true.[6] Yet, it is highly plausible, as both Sosa (2021) and Pritchard (2016) both agree (for different reasons), that the knowledge attained by A, whose inquiry is characterized by its independence from assistance from others, is, in some sense, epistemically *better* than the knowledge attained by B, whose inquiry boils down to trusting that whoever wrote the answer key was right. If this intuition can be vindicated by reference to the independence from others exhibited by A's method, then it would seem that negative autonomy is indeed a significant factor in epistemic axiology. Sosa and Pritchard provide sophisticated frameworks for articulating just what this "surplus value" in case A consists in, and in doing so, they inadvertently lend what looks (initially at least) like some compelling support to the NEA theses.

5.3 Sosa's Approach

Within Ernest Sosa's (2021) influential telic virtue epistemology, knowledge is understood as a kind of performance, and its value is assessed accordingly.[7] The central concept is that of an *apt belief*—a belief that is true *because* it is the product of competence. This framework distinguishes between mere "animal knowledge" (apt belief) and a higher grade of "reflective knowledge" or "knowledge full well," which involves not only an apt belief but also an apt grasp of the belief's aptness.[8] In *Epistemic Explanations*, Sosa applies this framework to the kind of contrast seen in our mathematician cases. For Sosa, mathematician A's success is not merely in acquiring a true belief but also in achieving *firsthand knowledge*, which is knowledge attained through the exercise of her own *firsthand competence*—in this case, her own reasoning faculties. By working through the proof herself, she gains what Sosa argues is essential for genuine *understanding-why*—namely, an "insight of one's own into the premises and into the immediate inferences that constitute a proof" (2021, 5). This is a form of knowledge based on *experiencing what makes the proposition true*, a "direct confrontation" with the reasons themselves.

Mathematician B, by contrast, will, for Sosa, fall short. Although she acquires knowledge through testimony,[9] she lacks this deeper good Sosa points to. She "falls short in h[er] understanding of why the theorem is true, lacking as [s]he does an adequate grasp of any proof" (2021, 4). Her success is a "mere success," something she could have achieved by peeking at the answer. Mathematician A, however, aimed for a more valuable goal: *firsthand success*, where the truth is attained through an apt performance attributable to her own competence. In doing so, she engages in an activity through which, on Sosa's view (2021, 6), a person "flourish[es] as a rational animal." By deferring, mathematician B "foolishly neglects the proper objective" of the exercise, which is to aim for the higher gnoseological good of firsthand success over the merely practical good of an efficient, true belief.

5.4 Pritchard's Approach

Duncan Pritchard's (2016) preferred framework (what he now calls "anti-risk virtue epistemology"; see Pritchard 2020) also provides tools to diagnose the value difference between A and B. For Pritchard, knowledge requires a true belief that is not only the product of the agent's cognitive ability but also safe from being wrong (i.e., in the sense of suitably risk delimiting). More to the point, Pritchard holds that certain epistemic states possess a special value because they are what he calls "strong cognitive achievements." A strong cognitive achievement is a cognitive success that is primarily creditable to the agent's cognitive agency and involves either the overcoming of a significant obstacle or the exercise of a significant level of skill. From this perspective, mathematician A's solitary effort is a paradigm case of such an achievement. Her success is attributable almost entirely to her own significant skill and hard work. Through this process, she comes to, to use Pritchard's preferred locution, "see for herself" that the theorem is true, engaging in a form of "intellectual seeing" that yields a "proper explanatory grip" on the matter. Pritchard argues that this grasp of explanatory connections is precisely what sheer deference (as we have in case B) can't provide. So, by proving the theorem herself, she takes "ownership of [her] epistemic position," which is a status unattainable by simply accepting the testimony of others. Mathematician B, while gaining knowledge (we're assuming again the textbook is reliable, there's minimal risk of error, given the way they consult the back pages), does not secure a strong cognitive achievement in this sense. Her success is creditable more to the textbook author than to herself; her knowledge is less valuable.

5.5 Analysis of Prima Facie Appeal

With Sosa's and Pritchard's accounts now on the table, the prima facie appeal of the negative autonomy theses becomes clear. Both identify a surplus of epistemic value present in case A and absent in case B. Sosa locates it in firsthand insight (achieved via exercise of firsthand competence) exhibited in the (intentional) pursuit of a higher epistemic goal than *mere* knowledge; Pritchard locates it in the attainment of a strong cognitive achievement whereby one "sees for themselves" that the proof is true. On both pictures, a salient difference between the two cases each use to illustrate a difference in epistemic value is the *degree of intellectual independence*. Put simply, mathematician A's inquiry looks autonomous; mathematician B's is dependent. It is therefore tempting to conclude that this independence *is itself the source of the surplus value*. Negative epistemic autonomy thus presents itself as the obvious axiological difference maker.

The power of this intuition, however, rests, we think, on the artificially polarized nature of the comparison. The thought here is as follows: When what's at issue is contrasting a case of what looks like near-total independence on others' assistance with one of near-total dependence, the variable of autonomy becomes perfectly correlated with the presence of certain other valuable epistemic goods such as understanding and explanatory grasp. This makes it easy to mistake this correlation for causation or even identity. The next section will challenge this illusion directly by introducing a case that "breaks" the correlation in such a way as to demonstrate that deep epistemic dependence is (despite initial appearances) fully compatible with the very epistemic goods that Sosa and Pritchard rightly champion.

5.6 A Mixed-Methods Case

Although it is true that mathematician A exhibits negative autonomy in her solitary pursuit of the proof, we contend that this independence is merely an incidental feature of the case. The negative autonomy present in case A does no genuine work in explaining the value difference between A and B.

This point is easily missed when the comparison is restricted to these two (near) limit cases. To break the illusion, let us introduce a third inquirer:

Case C (mixed-methods case): Mathematician C proves theorem T while: (i) using an AI assistant to check intermediate steps for accuracy, (ii) consulting with colleagues about different proof techniques, and (iii) relying on digital tools for complex calculations.

Case C is a paradigm of dependent inquiry—namely, relying on a variety of external aids, such as an AI assistant, colleagues, digital tools, and so forth. According to the logic of the NEA theses, the knowledge achieved by mathematician C must be epistemically inferior to that achieved by mathematician A.

But is it? We think not. Far from being an inferior achievement, mathematician C's success represents an equally, if not more, valuable epistemic achievement. Her method is not one of passive deference, like mathematician B, but rather of active, critical integration of information and sources. She exercises a sophisticated, higher-order skill—that is, the skill of, put roughly, managing and synthesizing diverse epistemic resources into a coherent proof. In doing so, she achieves—and indeed this is difficult to deny—precisely the kinds of epistemic goods that Sosa and Pritchard champion. She attains what Sosa calls "firsthand knowledge" because she makes the reasons her own; her use of testimony and tools is as a "conduit to reasons which can then be accessed firsthand" (2021, 14) rather than simply deferring, and her resulting grasp of the proof is anything but the "truncated understanding" Sosa attributes to mathematician B. Likewise, her success is a paradigm of what Pritchard calls a "strong cognitive achievement," one attributable to her skillful agency in overcoming a complex intellectual challenge. She takes "ownership of [her] epistemic position" *not by shunning dependence* but rather by, as we may make explicit in the example, mastering it for the task at hand. We emphasize here that this is much as, in actual research mathematics, Fields medalist Terence Tao increasingly uses LEAN automated proofing and AI regularly at the frontier of new discovery[10] and encourages others to get as skilled as they can at utilizing these research tools to get new results; Tao seeks deep understanding not testimony that proofs are correct.

The conclusion that case C is epistemically inferior to case A is false, and so the premise that led to it—that epistemic value decreases in proportion to epistemic dependence—must also be false. Negative autonomy cannot be the feature that explains the epistemic superiority of case A over case B. The value of A's performance must be rooted in something else entirely. This can be seen even more clearly by examining real-world cases of complex, collaborative inquiry.

5.7 Real-World Example: DNA Discovery

Case C is not contrived or idiosyncratic to mathematics. Let's look at two real-life example cases from science that help bring this out, beginning with a classic. The discovery of the double helix structure of DNA is a canonical example of a profound scientific achievement. Yet, it was an achievement born not of solitary genius but rather of the integration of disparate, dependent lines of inquiry—a paradigm case of distributed cognition (Palermos 2020; Kirchhoff and Newsome 2012). The final model proposed by James Watson and Francis Crick (hereafter, W&C) was the result of a (highly skilled) synthesis of data provided by others (most notably, Rosalind Franklin) often without their full consent or collaboration.

Consider here how W&C depended on other people using instruments and passing on the data. For one thing, the X-ray diffraction image known as "Photo 51" captured by Rosalind Franklin and her student Raymond Gosling provided important evidence for a helix structure (Maddox 2002). Here is the structure of the dependence here: The data was a product of both "instrumental dependence" (on the X-ray) and testimonial dependence (e.g., when Maurice Wilkins famously showed the photo to Watson without Franklin's permission). But there's also clearly dependence here, on the part of W&C, on prior discovery. They relied, and indispensably so, on Chargaff's "base-pairing" rules—namely that the amount of adenine equals the amount of thymine, the amount of guanine equals the amount of cytosine (Chargaff et al. 1951). The resulting understanding of DNA's structure that they attained was really a monumental cognitive achievement, but it was one that was deeply dependent on parts of the overall task being outsourced as well as to testimony from that outsourced work. Clearly no single individual possessed all the necessary evidence or skills. The "seeing for oneself" that Pritchard champions as well as the understanding-why that Sosa mentions (in firsthand inquiry) was attained here, just not in such a way that each step of the process was through the workings of a single, autonomous mind. The value of the achievement lay in large part in the successful integration of diverse cognitive resources, which is a process that the NEA theses would devalue due to their being riddled with dependence.

5.8 AI and AlphaFold

So, we've seen a historical case. Let's look now at a more contemporary one. As we see it, the rise of artificial intelligence in science (e.g., Erduran and Levrini 2024; Gao and Wang 2024) offers perhaps an even more straightforward challenge to the NEA framework. Without a doubt, we've seen how Google's DeepMind is good for more than playing amazing chess. DeepMind's AlphaFold has revolutionized fields such as structural biology by "protein folding" better than any human. This presents a unique epistemological scenario, one that one of us has discussed in detail in previous work (Carter 2024b).

On the one hand, AlphaFold, like other deep neural networks, operates as a "black box"; we just don't understand the higher-dimensional mathematics that accounts for how it gets things right. And yet, the proof is in the pudding. It generates highly reliable and even novel predictions; it's just that the principles or mechanisms by which it arrives at these predictions are largely inaccessible to human understanding. This is a well-known problem (which has led to some recent work on the epistemology of AI). Setting aside some of the tricky epistemological questions that emerge from thinking about testimony from black box deep neural networks, a point of agreement is on the descriptive claim that scientists are, increasingly, profoundly dependent on an opaque, nonhuman cognitive tool. Now, that said, scientists use AlphaFold's outputs as a foundation for further research, leading to breakthroughs in areas such as drug discovery. They achieve a deep understanding of the *implications* and *applications* of the predicted structures without having Sosa's "firsthand insight" into *how* those structures were derived—although, as a thought goes here, perhaps that is no more important than for W&C to have had "firsthand insight" into the earlier base-pairing rules that Chargaff and colleagues discovered, and which, for W&C's purposes, it was fine to just rely on (much as mathematicians don't prove from scratch each theorem they rely on for a more complex proof).

On a strict reading of $NEA_{evaluative}$, the knowledge gained from a tool such as AlphaFold should be cheap—a kind of "truncated understanding" on par with looking up an answer in a book. This view, however, fails to account for the facts. What we get from AlphaFold is the basis for high-stakes scientific discovery; we understand proteins better than we did before.

This presents a puzzle for NEA. Immensely valuable epistemic goods are being achieved through methods that are the very antithesis of intellectual independence. The achievement here is not the scientist's ability to replicate what the AI does. Rather, the achievement lies in the human scientist's skillful governance of the interaction between their own research goals and the AI's outputs. This looks like a higher-order skill of *managing* dependence, not *avoiding* it.

The cases of W&C and AlphaFold share a common pattern. The value we admire in solitary inquirers comes from their successful integration of multiple *internal* cognitive faculties. The value in collaborative and technologically mediated inquiry comes from the successful integration of multiple *external* cognitive resources. The underlying epistemic good is structurally analogous in both cases: The skillful integration of diverse inputs to solve a complex problem. The mistake made by champions of NEA theses is to confuse the *container* (solitary inquiry) with the valuable *content* (skillful integration).

5.9 A reductio

A separate line of argument exposes a serious flaw in the very structure of the negative autonomy theses. This argument takes the form of a reductio ad absurdum aimed at $NEA_{prescriptive}$. By demonstrating that the prescriptive norm overgeneralizes into absurdity, we expose the mistaken axiology of its evaluative counterpart, $NEA_{evaluative}$. The argument begins by challenging the artificially narrow concept of "dependence" that makes the solitary inquirer of case A appear to be a paragon of autonomy in the first place.

$NEA_{prescriptive}$ commands us, ceteris paribus, to prioritize the method of mathematician A over that of mathematician B. The grounds for this prescription are that A's inquiry is more independent than B's. Yet, mathematician A is only independent in a *very specific and limited sense*. To make this precise, we can introduce a distinction between two forms of epistemic dependence: synchronic epistemic dependence and diachronic epistemic dependence:

Synchronic epistemic dependence: Reliance on the epistemic labor of another agent *at the time of inquiry* (e.g., asking an expert at a given choice point, such that that labor is performed while one continues to inquire).

Diachronic epistemic dependence: Reliance on epistemic labor that has been completed in the past and inherited by the agent. (This includes tacit reliance on already-established theorems, conceptual frameworks, and, most fundamentally, language and symbolic notation.)

Mathematician B's reliance on the textbook is a clear case of synchronic dependence. Mathematician A's inquiry, although entirely free from this specific form of dependence, is (significantly) shaped by diachronic dependence. Her entire project is made possible by an enormous inheritance of concepts, axioms, techniques, theorems, and so forth developed by others, from Euclid's launch of number theory to her immediate predecessors. Her knowledge of how to do mathematical proof itself is a form of testimonial knowledge acquired through years of study with teachers or at least through reading books. She is, as the metaphor goes, "standing on the shoulders of giants."[11] Also, insofar as we're looking at things strictly, the very *language* that mathematician A relies on to structure her thoughts—the symbolic notation of mathematics—is a shared, inherited cultural technology; she didn't invent that notation. Drawing on the perspective of 4E cognition (although with no requirement that we accept it for present purposes), we can view this notation not merely as a passive record of thought but also as an external cognitive tool that is actively coupled with her brain to perform the reasoning itself (Menary 2010, 2012). On this way of thinking, her thinking is constitutively and not just causally dependent on this external artifact (Palermos 2014, 2015; Clark 2010; Adams and Aizawa 2008).

With the above in mind, the proponent of the NEA theses now faces a dilemma. The principle as stated makes no distinction between synchronic and diachronic dependence; dependence is dependence. Thus, either the norm applies only to synchronic dependence or it applies to all dependence.

Consider now "Horn 1": The norm applies only to synchronic dependence. The problem with going this route is that such a restriction looks ad hoc and unprincipled. What nonarbitrary reason would there be after all for valuing independence from a living colleague but not from a deceased one like Euclid, whose work is accessed via the testimony of a textbook or a remembered theorem? The distinction looks like it tracks nothing of deep epistemic importance.

So, this looks like a reason for the proponent of $NEA_{prescriptive}$ to go with "Horn 2" and say that the norm applies to all dependence. This horn, although more consistent, leads directly to a reductio. To see why, let's introduce a new character in variation on the case of mathematician A:

Case A*: Verificationist mathematician: Like mathematician A, mathematician A* works alone. However, she believes that relying on the proofs of others, such as Euclid, would render her too dependent. Therefore, A* refuses to rely on any previously established theorems she did not prove herself from axioms she has personally vetted. She attempts to prove theorem T (more or less) ab initio.

Now, compare case A with case A*. Mathematician A is dependent on the diachronic testimony of the mathematical community. Mathematician A* has intentionally eschewed this dependence. She is, therefore, more intellectually independent; she exhibits a higher degree of negative epistemic autonomy. If $NEA_{prescriptive}$ is correct, it must follow that, ceteris paribus, we ought to prioritize the method of A* over A.

But this is a bad result. This conclusion is already deeply implausible, as it would effectively recommend reinventing the wheel at every turn where this is available without knowledge loss. And yet, the reductio does not stop here. Even our radical mathematician, A*, is not fully autonomous. She still relies on the preexisting, socially generated tool of mathematical notation. To rely on this inherited symbolic system is to depend on an external cognitive artifact. Therefore, if $NEA_{prescriptive}$ is to be applied consistently, it must favor the approach of an even more radical inquirer, call them the "symbol creator":

Case A: The symbol creator:** This mathematician attempts to prove T not only ab initio but also without relying on conventional notation. She first invents her own unique symbolic language for expressing mathematical concepts before proceeding with her proof.

Here, the absurdity becomes manifest. $NEA_{prescriptive}$ would entail that, all else being equal (i.e., if we could answer our question knowledgeably either way), we ought to prioritize the intellectual path of A** over A*, and A* over A. The norm systematically recommends stripping away the essential scaffolding of collaborative human knowledge: not a promising idea. The ideal it points toward is not an epistemic exemplar but rather a thinker that is basically hamstrung by a pathological commitment to "intellectual autarky." We've seen how autarky-style economic nationalism works out politically; it is not a success.[12] The final turn of the screw is to recognize that even A** is dependent on the structure of natural language to formulate the concepts that precede her symbolic system. The ultimate endpoint of $NEA_{prescriptive}$ would be to prioritize, insofar as possible, inquiry free from the inherited structures of language itself—a basically incoherent goal.[13]

Since $NEA_{prescriptive}$ generates absurd prescriptions, it must be rejected as a normative principle. And as we see, it fails because it is based on a mistaken axiology. The extra epistemic value intuited in case A but not B cannot be explained in terms of the negative autonomy that features in the belief-forming process in A because a consistent application leads to ruin. The very things the norm would have us discard if we could—reliance on established knowledge, shared notation, and linguistic frameworks—are not epistemic liabilities. They feature as essential preconditions for most all meaningful intellectual achievements. This overgeneralization objection not only compels us to do away with the prescriptive norm but also, by extension, counts against the evaluative norm, $NEA_{evaluative}$, upon which it rests. In boilerplate form, if independence is not something we ought to prioritize, then it cannot be the source of epistemic value its proponents claim it is.

5.10 Taking Stock and Conclusion

Let's take stock. The foregoing arguments show that epistemic value does not robustly track first-order negative autonomy—independence simpliciter in the very process of belief formation. The mixed-methods case (C) demonstrated that deep dependence can coexist with understanding-why and strong cognitive achievement; and the reductio from $A \rightarrow A^* \rightarrow A^{**}$ showed that prioritizing independence for its own sake yields absurd prescriptions.

This section concludes by doing two things. First (section 5.10.1), it situates that result (and the more general result that we should reject the NEA norms) within a more general picture of autonomy—*relational autonomy*—and the analogue idea that epistemic autonomy (of the sort worth valuing as inquirers) is relational. Second (section 5.10.2), it clarifies the residual, instrumental role of certain forms of independence in information-dense, attention-hijacking environments of the sort discussed in chapter 2.

5.10.1 Relational Autonomy

A familiar thought, traceable to Aristotle, is that human flourishing is, in an important sense, relational, even though we explain its relational character by appeal to features of individual natures (Broadie 1991). Human beings are such that standing in the right social relations partly fixes what it is for them to live well. If we take this idea as a starting point, it begins to look strange to suppose doing things (be it acting or inquiring) in isolation should be valued for its own sake rather than, at most, as an occasional means (more on this shortly).

A parallel insight animates four decades of work on relational autonomy in moral and political philosophy (e.g., Christman 2004; Mackenzie 2014; Stoljar 2000; Walter and Ross 2014; Westlund 2009). On these views, agents are self-governing through appropriate social, material, and institutional relations—not *despite* them. As John Christman puts it: "'Relational autonomy' is the label that has been given to an alternative conception of what it means to be a free, self-governing agent who is also socially constituted and who possibly defines her basic value commitments in terms of interpersonal relations and mutual dependencies" (2004, 144). Against individualized (Marlboro Man style) pictures of autonomy that implicitly oppose autonomy to dependence (while assimilating autonomy to self-directedness), relational accounts hold that capacities for self-direction are paradigmatically developed and sustained by socialization and scaffolding (Mackenzie and Stoljar 2000; Nedelsky 1989; Oshana 1998, 2006), and that whether a given exercise of agency is autonomous depends in part on the *quality of the agent's standing relations*—recognition, non-domination, accountability, and dialogical answerability (Anderson and Honneth 2005; Westlund 2009; Nedelsky 2011) rather than on, say, the absence of the obtaining such relations or (assuming they are in place) on independence from reliance on such relations in the course of meeting our goals.

A full defense of relational autonomy in practical life is not our task here.[14] The point is just conditional. If relational autonomy captures broadly the right shape for practical agency, then we ought to expect as a default position (absent some reason not to) an epistemic analogue—namely, mutatis mutandis, valuable autonomy for belief and inquiry is something we should expect will be at least partly realized through the quality of one's *epistemic relations* with others[15]—for example, recognition, non-domination, accountability, dialogical answerability—rather than by the absence of reliance on any such relations. Put differently, if the kind of autonomy worth valuing in our practical lives is the kind that contributes to flourishing—and if such flourishing is itself constitutively social—then, by parity of reasoning, the same holds in the epistemic case. If epistemic autonomy worth valuing contributes to intellectual flourishing, and if that flourishing is likewise relational, then epistemic autonomy, like its practical counterpart, will plausibly be relational in character. It will often involve competently authorizing, monitoring, and revising our epistemic dependencies rather than seeking to secede from them.

Two implications follow from this analogical thinking from the practical case. First, the capacities and competences constitutive of epistemic *self*-governance (of the sort worth valuing as inquirers) aren't going to be "quarantine-like." They will need to include such things as calibrating trust, managing uptake of testimony, integrating heterogeneous evidence, knowing when (and to whom) to outsource subtasks (instruments, experts, or AI systems), identifying defeaters, revising dependence as circumstances shift, and so forth.

Second, whether one's believing, gathering evidence—in the sense relevant to credit, ownership, and understanding—is relevantly her "own" is something we'd typically work out (drawing again from our analogy to relational autonomy in practical life) by considering whether the dependencies one manages in the course of her self-government are (some combination of) (a) agent-authorized (the subject stands behind them as hers; Benson 2005; Westlund 2009), (b) reason answerable (the subject could cite and revise the grounds of uptake; Fischer and Ravizza 1998), and (c) relation apt (they obtain within structures of recognition and non-domination rather than distortion or coercion; Fricker 2007; Nedelsky 2011; Oshana 2006). This is, again, all just the default we should expect in the intellectual case from a starting point where we're assuming practical autonomy is, like flourishing, itself relational.

With all this in mind, let's now reread our mathematicians A, B, and C through this lens. What made A's result better than B's was never "bare independence." However, one thing we can say is that A (unlike B) effectively made the reasons her own. But C can do that while depending widely—by competently governing her dependencies, exactly what any extension of relational autonomy generally to autonomous inquiry is going to predict. B's shortcoming on this line of thinking is *not dependence per se* (as the NEA theses would have it) but more something more akin to a failure of governance that both A and C manage, albeit differently. She merely relies where she should, to a first approximation here, appropriate and integrate. Likewise, A* and A**, we can appreciate, are defective for the same reason that B is: They undermine (or cut themselves off from) the very social and material conditions that enable epistemic self-governance, for thinkers like us at least. A* and A** may get to the truth this way, but they would do so despite this error in governance.

This way of thinking preserves what Sosa and Pritchard were right to prize—credit, ownership, understanding-why—while at the same time opening up a way of seeing how those kinds of epistemic goods are such that their attainment is compatible with (and often requires) pervasive dependence. For example, understanding-why is typically secured when the agent has appropriate grip on the sources, status, and inferential roles of the reasons she integrates (something sheer secondhand deference typically won't afford one). Strong cognitive achievement (which Pritchard thinks is lacking in cases such as B) is present when success is attributable to the agent's skill, albeit this will (from a starting point where we take epistemic autonomy to be plausibly relational) include skill in organizing and managing a distributed epistemic process and so forth. On both fronts, then, we arrive at the idea that governance of epistemic dependence—not its minimization—is itself a species of epistemic self-governance.

5.10.2 Derivative Value of Independence

So, is there, then, any value left to recover under the banner of *independence*? Yes—but only of a derivative kind, one that shouldn't be conflated with genuine *epistemic* value.

Here it will be useful to take a step back. A common mistake in value theory is to infer from "X is routinely necessary for valuable Y" that "X itself has Y-type value." Money enables many goods without being itself a prudential or moral good of the same kind. A safety harness is crucial to a climber's success without being a mountaineering achievement (buying a safety harness on eBay from one's couch is not a *mountaineering* achievement). Not everyone in the group of people who can find good harness deals on eBay is a good mountaineer or even a mountaineer. Buying a nice calculator or even making one from scratch doesn't earn "calculus" credit and so forth.[16]

This point has important relevance when it comes to thinking about conditions that need to be obtained for us to have a good epistemic success rate when navigating information-dense, high-choice environments—for example, what Floridi (2014) calls the "infosphere" we nowadays find ourselves in when seeking out information, especially online. The story here is a familiar one already from chapter 2: Our information environments are increasingly engineered to capture and redirect attention (Lavie 2010). High-salience, low-quality inputs—notifications, outrage cues, novelty bait, and so forth (consider many of the things that pop up on your phone first thing in the morning and frame your intellectual responses, the course of your inquiries, etc.)[17]—distort uptake. They swamp working memory and, in doing so, have the effect of derailing inquiries you might already set for yourself and which reflect your actual values (Bawden and Robinson 2020; see also Alfano et al. 2018).

In such conditions, a modest, selective kind of independence—the ability to step back from hijacking cues, quarantine sources that routinely mislead, and structure one's attention—functions as an *enabler* of competent inquiry. Crucially, the value here (like in the money and mountaineering examples above) is derivative: Going offline, filtering inputs, or choosing to work solo for a phase of inquiry has epistemic significance only insofar as it protects or enhances the capacities that constitute intellectual success.

Just as good nutrition and sleep enable performance without actually constituting athletic performances, likewise, as we see it, in certain cases, environmental independence (something we might attain by, e.g., shifting our environment from informationally dense to informationally sparse, cutting ourselves off from testimony, foregoing availability of cognitive tools, etc.) is not an epistemic achievement in itself; it is not of any epistemic final value. Even more, when independence of this sort does no enabling work (e.g., when it is misguided)—when it amounts to stubbornness, needless reinvention, or indiscriminate suspicion (e.g., cases such as A* and A**)—it has no epistemic payoff at all, not even instrumental. When, on the other hand, such strategic independence *does* enable the successful exercise of epistemic competence by, for example, reducing distortion, shoring up attention, or preventing domination, and so forth, its value here is at best strictly of the enabling variety, of an ilk with (in the athletic case) good nutrition, sleep, and so forth.

5.10.3 Summary

Let's now draw the threads together. The A/B contrast—the stalking horse of this chapter—tempts us to equate independence with the competences that ground *understanding-why* and epistemic ownership, goods we rightly value for their own sake. But C-style cases break that link. They show that such competences can be exercised through extensive, well-governed dependence. And A* and A** show the opposite danger: Fetishizing independence yields prescriptions that sever the very scaffolding that makes human inquiry possible—pointlessly so, absent some independent interest in lone-wolf inquiry.

A better account, drawing on relational theories of autonomy outside epistemology, explains these cases more naturally. On this picture, epistemically autonomous thinkers steer their dependencies under reasons and within relations of recognition and non-domination, much as practically autonomous agents do—assuming, as argued, that practical autonomy is relational. Within that framework, we can locate a derivative value for independence: In information-dense, attention-hijacking environments, certain forms of independence are valuable not because they constitute epistemic achievements but rather because they enable them. By resisting attention capture, avoiding systematically distorting channels, or carving out solo work phases where group dynamics would bias uptake, independence becomes a tool for managing dependence, not an alternative to it.

In the next chapter, we approach information-dense environments from the *upstream* end of inquiry, asking how their choice architectures "nudge" our interrogative attitudes—what we become curious about—and what that means for the kind of autonomy applicable to how we seek out answers in inquiry: erotetic autonomy.

6 Nudging Inquiry: Erotetic Attitudes and Intellectual Autonomy

6.1 Introduction

Recent literature on epistemic nudging has so far focused almost exclusively on what can be termed "doxastic nudging": the practice of tweaking the context in which agents form beliefs to make it more likely they will arrive at beliefs that are in their own epistemic interest, such as true beliefs (Grundmann 2021, 2023; Piedrahita and Vermaire 2025; Fratantonio 2024; Ryan 2025; Miyazono 2023). The foundational idea of nudging, as introduced by Richard Thaler and Cass Sunstein (2003, 2008), involves altering the "choice architecture"—the context in which people make decisions—to steer them toward choices that benefit them, but without resorting to coercion or mandates. Just as one might be nudged toward choosing fruit over candy by placing the fruit at eye level (so it becomes more salient to the shopper), one might likewise be nudged toward believing a scientific consensus over a conspiracy theory through subtle changes in information presentation on a website.

However, a singular focus on specifically *belief*-directed influence overlooks a separate aspect of our intellectual lives subject to influence: how we ask questions. The recent "zetetic turn" in epistemology (Kelp 2020, 2021; J. Friedman 2013, 2020; Falbo 2021; Millson 2020; Willard-Kyle 2023), which we registered toward the end of the previous chapter, has drawn attention to the fact that our epistemic endeavors are not just about holding high-quality static beliefs but also include doing well when it comes to dynamic processes of *inquiry*. Inquiry, at its core, is an attempt to answer questions. As Kelp puts it, "When we are inquiring into a given question, we are trying to *settle* this question" (2020, 359). This process involves a cycle of both doxastic attitudes (e.g., belief and suspicion), which are directed at propositions, and *interrogative attitudes* (e.g., curiosity, wonder, and investigation), which are directed at questions (J. Friedman 2013). We wonder *whether p*, and this wondering motivates an inquiry that might culminate in a belief *that p*. The interrogative attitude is, in this respect, "upstream" from the doxastic one.

This chapter introduces and defends the concept of erotetic nudging—a form of epistemic nudging aimed not at our beliefs but rather at our interrogative attitudes. If, as this chapter will argue, our attitudes toward questions are themselves "nudgeable," then the prevailing association so far of epistemic nudging with doxastic nudging has suppressed an underexplored area of social-epistemic intervention. And the practical significance of this oversight matters. Put simply, in an era of rampant misinformation,[1] the problem often lies not just in the downstream acceptance of bad answers but also further upstream in the asking of bad questions. The digital information environments we inhabit, which are nowadays governed by predictive analytics and recommender systems, are powerful new forms of choice architecture that shape not only what we come to believe but also, more fundamentally, what we become curious about and how we express that curiosity. In light of this, understanding how these architectures can nudge our inquiries is therefore a task at least as relevant to epistemology as getting a grip on doxastic nudging.

To this end, we'll proceed as follows. Section 6.2 clarifies the very idea of nudging and, in doing so, will distinguish the *definition* of a nudge from the separate normative evaluation of its use—a distinction that matters for avoiding common confusions in the literature that have run these two things together. Section 6.3 will then develop a principled functional taxonomy of erotetic nudging that identifies and analyses three distinct mechanisms by which our interrogative attitudes can be influenced. Then, building on this descriptive account, section 6.4 will turn to the central normative question, which asks about the relationship between erotetic nudging and a distinct species of intellectual autonomy we can call (positive) "erotetic autonomy": the capacity for self-governance as an inquirer. It will be argued that erotetic nudging, far from being inherently manipulative or divorced from reasons we have altogether, can be understood as a practice of providing, building off an earlier framework due to Levy, us reasons to inquire, a practice that is often compatible with, and sometimes even promotional of, erotetic autonomy. Section 6.5 then addresses address pathological cases of online influence, such as the "radicalization pipeline" on platforms such as YouTube (Alfano et al. 2021) and argues that these represent a more pernicious form of influence that fails the definitional criteria for a nudge. Finally, section 6.6 concludes by looking at some practical and philosophical implications of this framework for designing epistemic environments that actually foster epistemically beneficial erotetic nudging.

6.2 Understanding Nudging

6.2.1 Definition Versus Normative Evaluation

Debates surrounding nudging—which have been taking place across different academic disciplines from health-care research to philosophy to climate and computer science and public policy (Thaler and Sunstein 2008; Blumenthal-Barby and Burroughs 2012; Bovens 2009; Hausman and Welch 2010; Joly Chock 2020; V. Kumar 2016; Parmer 2023)—are often marked by a conceptual confusion that risks muddying the waters. This is the conflation of the *definition* of what a nudge is with the distinct *ethical question* of whether a given nudge is morally permissible—in the sense that it sufficiently respects autonomy, is not "too manipulative" or "sneaky"[2] or, as Sunstein frames the worry, an "insult to human agency" (Sunstein 2018, 1).[3] To get the kind of foundation we'll need to think clearly about erotetic nudging, let's first separate these two issues.

A "nudge," according to the canonical definition provided by Thaler and Sunstein (2008), is "any aspect of the choice architecture that alters people's behavior in a predictable way without forbidding any options or significantly changing their economic incentives (p.6)." A crucial addendum to this definition is that "to count as a mere nudge, the intervention must be easy and cheap to avoid (p.6)." Banning junk food is a mandate, not a nudge; placing fruit at eye level is a nudge because the candy is still readily available to a determined consumer.

Yashar Saghai (2013) offers a more granular account that breaks down the structure of a nudge into four theoretically separable components. For any influence to count as a nudge at all, it must involve:

1. Tweaks to choice architecture: A modification of the context in which a choice is made.
2. Triggering of shallow cognitive processes: The influence operates by leveraging predictable features of human psychology, such as cognitive biases and heuristics (often associated with "System 1" thinking) rather than engaging in explicit, deliberative reasoning ("System 2" thinking).

3. Preservation of the choice-set: The nudge must not eliminate options available to the agent; the set of choices must be unaltered or expanded compared with the baseline situation (Saghai 2013, 488).
4. Substantial non-control: The influence must be easily (enough) resistible. So, this condition means that the agent retains a meaningful degree of freedom and control over the outcome (i.e., you aren't nudged to do what you're compelled to do).

These four conditions provide a working template definition of what a nudge *is*, where, of course, there can be room for dispute about how to unpack specific details of any of these template components (see Piedrahita and Vermaire 2025). Although conditions 3 and 4 are designed to build into the concept of a nudge some minimal freedom-preserving constraints, they do not in and of themselves settle the deeper normative questions that animate critics of nudging. After all, worries about whether nudges are objectionably manipulative, whether they go too far in disrespecting agents' rationality, or whether they undermine autonomy are best understood as further evaluative questions to be asked *about* interventions that meet this structural definition. So, an intervention should, in principle, be able to satisfy all four definitional criteria and still be ethically problematic on other grounds (e.g., if it nudges people toward a harmful outcome, etc.). By keeping the definitional and the normative questions separate, the plan here will be to (i) first identify the mechanisms of erotetic nudging in a descriptive mode before (ii) turning to a focused evaluation of their impact on autonomy.

6.2.2 Doxastic Nudging

Proponents of doxastic nudging, as explored by, for example, Grundmann (2023) and others, aim to apply the established principles of nudging (for actions, as theorized initially about in literature on libertarian paternalism) to better understand how we can (suitably non-manipulatively) shape the formation of beliefs by choice architecture tweaks. The underlying premise here is straightforward: If certain actions, such as consuming unhealthy food, are not in an agent's practical self-interest, then equally, certain beliefs—for instance, vaccine skepticism or climate change denial—are not in their epistemic self-interest. Doxastic nudging thus investigates how shallow cognitive mechanisms might be exploited to facilitate epistemic objectives, notably the acquisition of true beliefs and the avoidance of false ones, of knowledge rather than ignorance and so forth.

This kind of "conceptual migration" (from political philosophy and public policy over to epistemology) requires a reinterpretation of the nudge's fundamental components so that they make sense within an epistemic context. What was once the "choice architecture" now translates to something like the evidential environment, and the "choice" itself becomes the adoption of a doxastic attitude. This translation, however, is not without its complications (Piedrahita and Vermaire 2025; Matheson and Joly Chock 2021). The very notions of "choice" and "freedom" prove considerably more intricate when applied to beliefs than to actions, and this is particularly so in light of the widespread philosophical rejection of a simplistic voluntarism of belief (Audi 1999; Vitz 2008). A principal concern, one that's been articulated by Grundmann (2021), is that doxastic nudges, by operating through *nonrational* mechanisms, may engender beliefs that lack an epistemically proper basis, thereby failing to qualify as justified or as knowledge. (An implication here that's bad is that "nudges" would, if this is right, succeed in, at most, steering people to unjustified true beliefs). In response to this line, proponents of doxastic nudging have mounted defenses for the epistemic credentials of beliefs that have been nudged.

6.2.3 Extending to Erotetic Nudging

Let's now begin to consider a further extension of the nudge framework: from the doxastic to the *erotetic*. The zetetic turn has illuminated inquiry as a process governed by *interrogative attitudes* such as curiosity and wonder. (J. Friedman 2013), attitudes whose objects are questions. So, should these very attitudes prove amenable to nudging, we would thereby be picking out a new and distinct domain of epistemic influence.

What we can call hereafter "erotetic nudging," then, is just precisely this practice. To a first approximation for now, erotetic nudging is the systematic adjustment of inquiry's choice architecture, calibrated to predictably modify an agent's *interrogative attitudes* in their interest. Table 6.1 shows the parallel structure that motivates this expansion and which positions erotetic nudging alongside its actional and doxastic analogues.

With a working account of erotetic nudging on the table, the rest of this chapter is organized around two aims. First, it answers a descriptive

Table 6.1
Nudging across domains

Domain of Influence	Target Attitude Type	Object of Attitude	Nudge Category
Practical	Intentions, desires	Action (e.g., choosing fruit)	Actional nudging
Doxastic	Beliefs, suspicions	Proposition (e.g., *that p*)	Doxastic nudging
Erotetic	Curiosity, wonder	Question (e.g., *whether p*)	Erotetic nudging

question: What are the specific mechanisms of erotetic nudging? We'll look at how the choice architecture of inquiry can be changed to influence interrogative attitudes through shallow cognitive processes. Second, it answers the normative question: When is this influence objectionably manipulative or a threat to the inquirer's autonomy? This issue parallels similar high-stakes problems for actional and doxastic nudging. The next section takes up the first question. The ones that follow address the second.

6.3 A Functional Taxonomy of Erotetic Nudging

This section organizes erotetic nudges by their *function* in inquiry. We distinguish between two functional types: (1) nudges that *initiate* inquiry and (2) nudges that *direct* an ongoing inquiry.

6.3.1 Generative Erotetic Nudging

The first of these two types we can call "generative erotetic nudging." Its function is to spark curiosity where there was none before. A clear example is *The New York Times* column, "What's Going on in this Picture?" An image of 488 Golden Retrievers in a field defies easy explanation (figure 6.1). It presents a cognitive gap that makes a viewer wonder, "Why are all those dogs there?" The nudge (the placement of the inexplicable picture itself) works by presenting a stimulus that gives the viewer a reason to ask "why?"

Generative nudges go beyond mere "strange question-prompting distractions" like the above picture. A second example is the Duncker Candle Problem,[4] known for illustrating a classic cognitive bias called "functional fixedness" (Adamson 1952)—although it can easily be redescribed in a way that highlights how we might nudge for the generation of interrogative attitudes. The functional fixedness bias is the bias of (in short) cognitive

Figure 6.1
488 Golden Retrievers in a field in Scotland. Source: Roddy Mackay for *The New York Times*. See https://www.nytimes.com/2024/01/04/learning/whats-going-on-in-this-picture-jan-8-2024.html.

inflexibility when it comes to thinking of objects as useful for purposes other than their typical functional purpose. The Duncker Candle Problem exploits this bias. Participants are asked to affix a candle on a cork board in such a way that the wax won't drip onto the table below, while using nothing but a book of matches and a box of thumbtacks. Some participants are presented with the box holding the tacks; others are presented with the tacks outside of the box. Most participants fail to solve it because of functional fixedness. They see the box of tacks only as a container. The question of whether the box could be a shelf just does not occur to them.

However, when the tacks are presented *outside* the box, people are far more likely to find the solution. Such placement operates as a generative erotetic nudge. By changing the presentation of the items, the choice architect makes it more likely the participant will ask the crucial question: "What *else* (besides holding tacks) can I do with this box?" Note that the nudge here does not exploit the bias directly. Rather, it works by circumventing the cognitive block imposed by functional fixedness in a way that "frees" the participant to inquire into new possibilities.

In both these above cases of generative erotetic nudging, notice that one is nudged to host an interrogative attitude (toward a question or a question type) *even when one lacked any antecedent curiosity* in that question or question type (or even general subject matter) prior to the nudge. In this respect, the nudging is generative of the interrogative attitude.

6.3.2 Directive Erotetic Nudging

A second functional category, directive erotetic nudging, applies to cases where an inquirer already possesses a general interrogative attitude (e.g., is curious about a topic), and the nudge then functions to shape or steer that inquiry toward a more specific question or a particular kind of answer. This can happen in at least two distinct ways: (i) by introducing presuppositions or (ii) by framing the space of possible answers.

6.3.2.1 Presuppositional erotetic nudging

What we can call "presuppositional erotetic nudging" is a form of directive nudging that steers an ongoing inquiry by subtly introducing semantic presuppositions that constrain the questions an agent is likely to ask next. A presupposition is an implicit assumption whose truth is taken for granted in an utterance. Granted, not all questions carry substantive presuppositions (e.g., "Is there a God?"). Many do, especially the *wh*-questions that are ubiquitous in online search environments. For example, asking something simple like "Who is the king of France?" presupposes that there is a king of France.

Digital choice architectures, such as search engine autocompletion, provide a powerful mechanism for this kind of nudging. Just consider an agent who is already curious about elite guitar solos and begins a Google search with the incomplete query "Is the best guitar solo by. . . ." Before they can finish typing their intended subject (e.g., "Jimi Hendrix"), the predictive algorithm presents a ranked list of suggestions: "Prince," "Slash," "Pink Floyd." These suggestions are not neutral. They function as invitations to adopt a certain set of presuppositions. The nudge, to be clear, does not force the user to presuppose that "Prince is responsible for the best guitar solo." Rather, it works through salience bias, making Prince a highly prominent candidate and then implicitly inviting the user to adopt the weaker, more plausible presupposition that "Prince is a live candidate for having the best solo." By accepting this (perhaps by clicking the suggestion), the user's inquiry is directed. Their next question might become "What is the best Pink Floyd guitar solo?," a question they might never have formulated otherwise. The nudge has successfully directed their existing curiosity (in guitar solos) by tweaking the salience of potential presuppositions within their inquiry's choice architecture, a process that triggers shallow cognitive processes responsive to what is presented first and foremost.

6.3.2.2 Erotetic nudging by partition-framing

A second, distinct way to direct inquiry is by influencing how an agent conceives of the space of possible answers to their question. This mechanism (for lack of a better term), erotetic nudging by partition-framing, can be defended against the charge of triviality by drawing a careful distinction between an *interrogative sentence* and the specific *semantic question* it expresses. In formal semantics, a question is often identified with a *partition* over a set of possible worlds, where each cell in the partition corresponds to a complete and exhaustive answer. A single interrogative sentence, however, can be ambiguous, corresponding to multiple distinct partitions.

Consider an example involving Florence's Uffizi Gallery. The sentence (used by a gallery goer) "Where in the Uffizi is Botticelli's *Primavera* displayed?" is underspecified. It could be used to ask the question whose answers partition the possibilities by *floor* ({on the first floor, on the second floor . . .}) or the question whose answers partition the possibilities by *hall* ({in halls 1–5, in halls 6–9 . . .}). These are different semantic questions, seeking different levels of granularity for an answer. A partition-framing nudge works by making one of these potential partitions more salient than another and in such a way as to nudge the inquirer to resolve the ambiguity of their own utterance in a particular way.

Suppose a choice architect already knows that an inquirer is subject to the anchoring effect (Tversky and Kahneman 1974), a bias where initial information serves as a reference point for subsequent judgments.[5] If the architect first presents information about other artworks in the Uffizi by referencing their *floor*, this "floor-based" framing acts as an anchor. The inquirer is then more likely to conceptualize the answer to their own question in terms of floors and to feel their inquiry is resolved upon learning "It's on the second floor." They'll be more likely to view *that* as the answer to "Where in the Uffizi is Botticelli's *Primavera* displayed?" and terminate their inquiry. Now, had the architect anchored the context with "hall-based" descriptions, the inquirer would likely have sought a hall number and considered "the second floor" as just an incomplete, partial answer.

This mechanism at play here in cases of erotetic nudging by partition-framing is powerful because it influences not just the path of inquiry but also the inquirer's own standards for what counts as *closure of inquiry*. Consider a related example here for how this works, using not just locational granularity but also *depth/shallowness of description* as what's ambiguous between multiple distinct partitions. Suppose an inquirer might be perfectly satisfied with a general description of the figure-skating jump called a "Salchow" and may not desire (or even benefit from) a highly technical, fine-grain answer from biomechanics.[6] But partition-framing nudges can steer us toward seeking answers at a level of granularity that is more (or less) beneficial for *our* epistemic goals. The detective, for instance, investigating an injury from a fall during a Salchow jump, faces an ambiguous question: "Why did the skater fall?" This question could be answered by partitioning the possibilities along general lines of skater error ("because she lost her balance") or along fine-grain biomechanical lines ("because the boot's stiffness impeded the necessary ankle flexion for the jump's takeoff edge").

A choice architect's goals will determine which partition they nudge the detective toward. A lawyer suing the skate company wants the detective to pursue the fine-grain biomechanical question. They might nudge the detective by first presenting reports on equipment failures in other sports and so forth, anchoring the conversation in technical details. The detective is now more likely to seek a biomechanical cause and will consider a simple answer such as "she lost her balance" as being incomplete. But a lawyer for the skate company wants to focus on the skater's error obviously. They might nudge the detective toward the general partition by showing videos of the skater's past technical mistakes. By anchoring the inquiry in terms of "performance" and "form," the detective is now more likely to accept "poor technique" as a complete answer (bad for the lawyer suing the skate company), and may never even formulate the more specific question about the skate's design.

With the above in view, table 6.2 summarizes this functional taxonomy of erotetic nudging. The taxonomy, we should note, is not intended to be exhaustive; rather, it captures the main kinds of mechanisms that, for present purposes, are most relevant to understanding how interrogative attitudes can be influenced.

6.4 Erotetic Nudging and Autonomy

Having provided a working descriptive taxonomy of erotetic nudging, we can now turn to the central *normative question*: Does this practice objectionably

Table 6.2
A functional taxonomy of erotetic nudging

Nudge Type	Function in Inquiry	Key Mechanism	Cognitive Bias Leveraged	Example
Generative nudging	Initiating inquiry	Presenting a cognitive gap or anomaly; circumventing a cognitive block	Curiosity from anomaly; overcoming functional fixedness	*NYT* "What's Going on in this Picture?"; Duncker Candle Problem
Directive nudging (via presupposition)	Directing inquiry	Making certain presuppositions salient to constrain subsequent queries	Salience bias; availability heuristic	Google autocomplete suggesting specific artists for "best guitar solo"
Directive nudging (via partition-framing)	Directing inquiry	Making a particular partition of the answer space salient to frame the question's granularity	Anchoring effect	Uffizi Gallery example (framing answers by floor vs. by hall)

manipulate inquirers or undermine their autonomy? This section argues that, when properly constrained, erotetic nudging is not only compatible with but can even be promotional of *erotetic autonomy*—the capacity for intellectual self-governance in the domain of inquiry.

6.4.1 Conditions for Preserving Erotetic Autonomy

For an initial starting point, let's, say, think of (in broad template terms) erotetic autonomy as disposition to exercise self-governance when asking questions, seeking answers, and manifesting interrogative attitudes such as curiosity and wonder. Although a full theory of this capacity is beyond the scope of this chapter, we can identify for our purposes now at least two necessary conditions for its preservation in the face of various sorts of influence, and we can do this in a principled way by drawing on the definitional constraints established in section 6.2.1.

First, corresponding to the third definitional criterion of a nudge is a *choice-set preservation condition*. The distinction between tweaking the *context* versus the *content* of choice architecture is best understood as a requirement to preserve the agent's set of options. An influence attempt that actually *removes* questions from an agent's consideration (e.g., in an egregious case, by installing a virus that literally deletes certain autocomplete suggestions) would (as we're thinking about it) alter the content of the choice set and thus fail this condition (and, in so doing, cross from a nudge into a more coercive intervention). That said, note that the three species of erotetic nudging described in section 6.3 are, by design, meant to satisfy this condition. They alter the *presentation* of the inquiry's context, not the underlying set of questions an agent is actually free to pursue.

Second, corresponding to the fourth definitional criterion is a condition of *substantial non-control*. The influence must be "easy and cheap to avoid" (Thaler and Sunstein 2008). In the context of action, this is often cashed out in terms of easy resistibility (Saghai 2013). However, for mental states such as belief, something more specific to inquiry is needed. As Grundmann argues, doxastic control is not about the ability to choose beliefs at will (see also Vitz 2008; Audi 1999; Vermaire 2022) but rather about the agent's retained ability to *revise* those beliefs in light of new evidence.

A doxastic nudge is non-controlling if it "does not isolate doxastic states from rational influences" (Grundmann 2021).[7] This kind of model, which Grundmann suggests in the case of belief, provides a good analogue for the erotetic domain. Questions, like beliefs, are *revisable*. Indeed, successful inquiry often consists in a process of refining and revising one's questions. We can therefore unpack the non-control condition for erotetic nudging in terms of preserving *erotetic revisability*. The idea, then, is as follows: An erotetic nudge is substantially non-controlling if it does not undermine the inquirer's ability to rationally revise, reject, or refine the questions they are nudged toward.

Contrast a good and a bad case vis-à-vis the substantial non-controlling condition. Consider an online conspiracy forum. A new member asking critical questions (e.g., "What is the primary source for this claim?"—e.g., that Pizzagate is true) is met not with evidence but rather with *social punishment*—they might be labeled a "shill" or a "sheep." This environment is engineered to make the rational revision of one's questions psychologically costly. The influence fails the "easy and cheap to avoid" test because rejecting the group's accepted questions risks social exclusion. The user's ability to revise their inquiry is thereby undermined. In contrast (and here's a good case), consider a restaurant menu that lists the chef's specials in a prominent box. This design nudges the diner to inquire about those specific dishes ("How is the sea bass prepared?"). Here the diner's ability to reject or refine this question is not at all undermined. They can easily just ignore the specials and ask about the pasta section instead. The nudge is a resistible suggestion that fully preserves the diner's erotetic revisability.

6.4.2 Nudging as Providing Reasons to Inquire

The most persistent objection to nudging, generally speaking, is that by triggering "shallow" or non-deliberative cognitive processes, it bypasses our rational capacities and is therefore manipulative. However, this objection rests on an overly narrow conception of what it is to be rational and responsive to reasons. Building on recent work by Levy (2019b), this section advances the central positive thesis of the chapter: Well-designed erotetic nudges do not merely bypass reason but can also function as a way of providing *reasons to inquire*.

Levy's tack is to challenge the premise, common to both critics and proponents of nudging, that influence via shallow (type 1) processes is necessarily nonrational. The idea here is that many of our cognitive systems are designed to be reliably responsive to such reasons as favoring considerations in a fast, frugal, and unreflective way. So, as the thought goes, it's a mistake to equate "shallow processing" with "nonrational processing," even if shallow processing really is in tension with the species of rational processing we think of as deliberate, reflective, rational processing.

Once this broader view of reasons responsiveness is adopted, the mechanisms of erotetic nudging can be potentially redescribed in a new light. They are not, as it were, simply "sub-rational pokes." They are ways of presenting considerations that function as reasons for (or against) taking up, sustaining or refining interrogative attitudes. This connection can be made systematic by revisiting the taxonomy from section 6.3: *Generative nudges* provide the inquirer with reasons to open an inquiry. The anomalous *New York Times* photograph does not just cause curiosity; rather, it presents a powerful reason for it—an unexplained phenomenon demanding investigation. *Presuppositional nudges* provide the inquirer with reasons to adopt a certain presupposition to guide their inquiry. The salience of "Prince" in a Google autocomplete list provides a (weak, defeasible) reason to treat Prince as a relevant candidate, thereby warranting a more specific, presupposition-laden follow-up question. *Partition-framing nudges* provide the inquirer with reasons to adopt a certain standard of resolution for their inquiry. The "floor-based" framing in the Uffizi example provides an anchor that functions as a reason to adopt that level of granularity as sufficient for closing the question.

On this view, then, erotetic nudging is a practice that furnishes us with reasons to inquire in the first place, to predicate our inquiries on certain assumptions, and to seek answers of a particular kind. Rather than being inherently threatening to erotetic self-governance, this practice can be seen as a way of rationally guiding the process of inquiry. That is, by providing reasons—even reasons apprehended by our unreflective cognitive systems—erotetic nudging can engage rather than bypass our rational agency, broadly construed, and in a way that preserves and, in some cases, even enhances (particularly in cases of benevolent erotetic nudges as a result of epistemically benevolent choice architecture design) our capacity for self-governance as inquirers.

6.5 Responding to the Radicalization Pipeline Objection

Let's look now at an objection to what is so far an optimistic picture of erotetic nudging. What about the well-documented cases of online platforms seemingly leading users down "rabbit holes" of extremist or conspiratorial content? For example, one of us has, as a coauthor on a study by Alfano et al. (2021), suggested that YouTube's recommender system can nudge users who start with innocuous queries toward increasingly radical content—namely, a phenomenon sometimes called the "radicalization pipeline." Do such cases not simply prove that erotetic nudging, at least as it features in online choice architecture, is a dangerously manipulative tool? Let's look closely at this worry.

As we see it, the response to this objection requires returning to the template definitional boundaries of a nudge. A core feature of a nudge, as distinct from more coercive forms of influence that are not genuine nudges but something else, is the substantial non-control condition: It must be sufficiently "easy and cheap to avoid." The pathological cases of online influence are best understood not as examples of genuine nudging but rather as instances of a more aggressive form of manipulation that *fails* this easy resistibility test. When an environment is designed in such a way that resisting its influence becomes (beyond some threshold) psychologically difficult, it crosses the line from nudging into something that looks more akin to coercion or "addiction-by-design."

To understand *why* certain digital environments can be so difficult to resist, the analysis of technologist and design ethicist Tristan Harris (2019) is instructive. Harris argues that many popular apps and platforms are designed like "slot machines" to "hijack our psychological vulnerabilities." The key mechanism behind the addictive power of a slot machine is not the reward itself but rather the *pattern* of the reward: *intermittent variable reinforcement*. The unpredictability of the reward (you pull the lever, and you might get a prize, or you might get nothing; e.g., Clark and Zack 2023; Kuem et al. 2021) is maximally compelling and drives repetitive, habitual behavior by producing the same neural changes seen in substance addiction (D. Ross et al. 2012).

Harris points out that this same reward structure is built into the core of many digital interfaces (for a wider discussion on this point, see Alter 2018). When we pull down to refresh our social media feed, we are basically pulling the lever of a slot machine to see what new content we get. When we check our phone for notifications, we are playing to see what social approval (likes, comments, messages) we have won. This design creates a cycle of craving and checking that can be extremely difficult for many users to break, where the goal is to maximize "time on site" for economic reasons (e.g., serving more advertisements), and this goal is achieved by creating a psychologically compelling, hard-to-resist experience.

We emphasize this kind of addiction structure because, as we see it, when a choice architecture is deliberately engineered to exploit these deep-seated psychological vulnerabilities using mechanisms such as intermittent variable rewards, *it is no longer a mere nudge*. Its influence is not "easy and cheap to avoid." It is a "hijack."[8] The existence of such manipulative designs does not, therefore, serve as a counterexample to the concept of benign erotetic nudging. Instead, it highlights the importance of the definitional boundary. The point is not that erotetic nudges are somehow uniquely immune to being hard to resist. Rather, it is that, *by definition*, an influence that is not easily resistible is not a nudge. The "radicalization pipeline" and "slot machine" designs are examples of a different and more pernicious kind of influence. Recognizing this allows us to isolate the phenomenon of genuine erotetic nudging—which is constrained by (an inquiry-relevant analogue of) the condition of *easy resistibility* in the case of actional nudges—and evaluate it on its own terms. Although some individuals may be more susceptible to even benign nudges, the paradigmatic cases of erotetic nudging discussed in this chapter are those that preserve the agent's ability to easily go their own way and can thus be interpreted as providing resistible reasons to inquire.

6.6 Conclusion and Implications

This chapter has sought to open up a new area of inquiry within the philosophy of epistemic nudging. By shifting the focus from nudging doxastic attitudes to the overlooked domain of interrogative attitudes, we've identified and then gone some way toward analyzing the concept of erotetic nudging. This proceeded in three main stages. First, we separated the structure of a nudge from its normative evaluation. Second, we proposed a principled, functional taxonomy of erotetic nudging, which allowed us to distinguish between *generative* nudges that initiate inquiry and *directive* nudges that shape its path, via either presupposition or partition-framing. Third, we defended this practice against the charge of manipulation by arguing that well-designed erotetic nudges are compatible with, and can even promote, erotetic autonomy. This defense was grounded in the idea that such nudges function not by bypassing reason but rather by providing us with resistible *reasons to inquire* in epistemically beneficial ways.

If, as is increasingly clear, the spread of misinformation and conspiratorial thinking is fueled not only by bad beliefs but also by bad questions, then interventions aimed solely at debunking false claims are fighting only half the battle.[9] Erotetic nudging offers a potentially powerful and autonomy-preserving strategy for intervening further upstream in the process of inquiry. By subtly tweaking the choice architectures of our digital and physical environments, we can, in principle, furnish individuals with reasons to pursue more fruitful lines of investigation and to avoid chasing false or epistemically toxic leads. An online user searching for information on homeopathic remedies, for example, is asking a question whose answers are unlikely to be epistemically beneficial. They shouldn't ask that question just like, in the store, they shouldn't pick the cigarettes. A well-designed erotetic nudge could (like well-designed health nudges) furnish that user with a reason to inquire differently—for instance, by making salient a question about the scientific evidence for homeopathy's efficacy.

7 Epistemic Autonomy in a Loop-Ridden Landscape: Navigating Social-Epistemic Dependencies

7.1 Introduction

As inquirers who increasingly offload cognitive tasks to technology and to others, we are constantly at risk of getting swept up in various kinds of feedback loops. By "feedback loop," we mean a dependence relation that (i) involves tight, ongoing reciprocal coupling between the agent and the environment; (ii) exhibits a "ratcheting" dynamic (think: uptake today increases exposure/credence tomorrow); (iii) filters out or discredits corrective evidence (e.g., by selective visibility or coverage failures); and (iv) imposes exit costs that make disengagement harder than continuation. Dependence simpliciter (say, ordinary uptake of testimony from a competent source in a transparent setting) does not, by itself, instantiate a feedback loop. Our target is therefore dependence under specific coupling-and-filtering conditions, not deference as such.

That said, consider that our (now ubiquitous) reliance on external supports—a cornerstone of contemporary social epistemology, psychology, and cognitive science[1]—stands in tension with traditional, more individualistic conceptions of knowledge acquisition, including some of those canvassed in chapter 5. This tension generates a kind of catch-22, which this chapter sets out to explore—that is, some of the very social and technological scaffolding we use to enhance our epistemic reach at the same time exposes us to complex feedback dynamics that put us at risk of getting trapped in a performative loop that has nothing to do with acting and thinking in line with our real interests. In short, the tools we nowadays need to use for extending knowledge can, under certain conditions, become traps.

This predicament reveals the inadequacy of two naive responses. The first is a retreat to a hyper-individualistic ideal of epistemic independence, which several chapters have already, in short, challenged as problematic. The second naive response is an uncritical embrace of epistemic dependence that too dismissively risks manipulation and entrapment.

The feedback loops that pose a real threat to our capacity to self-govern our lives intellectually are not alien or otherwise weird intrusions into our epistemic lives. They are often emergent properties of very common social-epistemic heuristics that make us successful knowers. Mechanisms such as trust in reliable sources, deference to consensus, and reliance on the informational coverage of our community are all rational, effective strategies for navigating a high-density epistemic environment structured around epistemic division of labor. The danger arises when the *structure* of the epistemic environment—such as the manipulative design of an echo chamber or the calculated ignorance within a fraudulent network—hijacks these rational processes and turns them against the agent. The challenge of retaining epistemic self-governance in modern information ecosystems, therefore, is the challenge of figuring out not simply how to be "more rational" in a classical, individualistic sense, but also how to develop a higher-order sensitivity to the systemic contexts in which our first-order rational processes operate. Or so we will argue.

In what follows, we treat well-governed testimonial dependence as the baseline case. We reserve "loop" talk for cases in which dependence is embedded within architectures that supply the ratcheting and screening features just noted (evidence-resistant echo chambers, unenlightened star networks, algorithmic amplification, etc.). The view eventually reached will be to countenance a species of positive epistemic autonomy worth valuing under the description of a "capacity for adaptive control." This capacity, it will be argued, is grounded in a specific cognitive achievement, *practical* understanding, which enables an agent to skillfully navigate the "loop-ridden" epistemic landscapes in such a way as to harness beneficial dependencies while avoiding pernicious ones.

7.2 Mapping the Loop-Ridden Landscape

To map this tricky landscape (in terms of the features that will be relevant in what follows), it will be useful to borrow the language of dynamical systems theory, which studies how complex systems behave over time. This framework allows us to help ourselves to some more precise terminology for thinking about forces that shape our beliefs through "pushes" and "pulls" such that we can do so by going beyond loose metaphor. We'll further distinguish between environmental loops we encounter (already up and running) and generative loops we are responsible for creating ourselves, with the former being further divisible into those driven by *attractors* and those driven by *repellers*. A brief note on scope: The taxonomy laid out in this section describes structures that can arise out of dependence but are not entailed by it. Many mundane dependencies (e.g., single use of a calculator, isolated secondhand testimony, brief exchanges, etc.) never realize these dynamics.

7.2.1 Attractor-Driven Loops

Within dynamical systems theory, an *attractor* is a state or set of states toward which a system tends to evolve, regardless of its precise starting point. An attractor creates a "basin of attraction." Once an object enters this basin, it is pulled toward the attractor, and escaping becomes progressively more difficult. A massive star's gravitational field is an illustrative analogy: Objects within its field are drawn inexorably toward it (figure 7.1).[2]

In the social-epistemic domain, certain structures function as powerful attractors for our beliefs and more broadly our allegiances. Depending on what we're being pulled toward, this could be good or bad.

An echo chamber is a prime example of an attractor-driven loop. It is crucial to distinguish it from what C. Thi. Nguyen (2020) terms an "epistemic

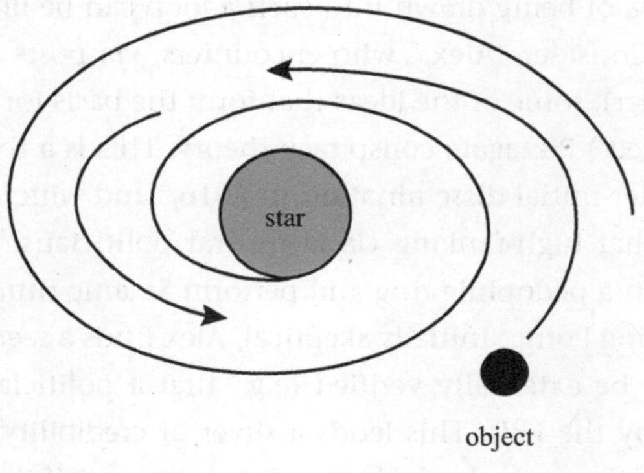

Figure 7.1
Gravitational field around a star.

bubble." An epistemic bubble is an impoverished (or at least limited or screened off) information network where dissenting voices have been omitted, perhaps accidentally. Qua bubble, it can be "burst" by simple exposure to new information. An echo chamber, on the other hand, is a social-epistemic structure where outside voices are not merely absent but are actively and systematically discredited (Begby 2024; Cinelli et al. 2021; Cookson et al. 2023; Elzinga 2022; Fantl 2021). This active discrediting is what gives the echo chamber its powerful, self-reinforcing "gravitational" pull.

Three structural features, adapted from Nguyen's (2020) well-known account, both define an echo chamber and help explain its function as a kind of dynamic attractor:

Doxastic exclusion: The group is composed of like-minded thinkers who share a core set of beliefs (let us say, beliefs centered on proposition *p*).

Discredit: Proponents of the contrary view (not-*p*) are systematically discredited. Their motives are impugned, their credentials questioned, and their arguments dismissed as malicious or misguided.[3]

Defeater reinforcement: Evidence that appears to count against the dominant viewpoint, *p*, is reinterpreted by the group as evidence for *p*. An attempt by a mainstream news source to debunk a conspiracy theory, for example, is framed as proof of that news source's complicity in the cover-up, thereby reinforcing the conspiracy theory itself.[4]

This third feature, defeater reinforcement, is the key to the attractor's robustness. It transforms potential escape routes into mechanisms that actually pull an individual deeper into the basin of attraction.

The very idea of being drawn into such a loop can be illustrated with a fictional case. Consider "Alex," who encounters, via posts on X (formerly known as Twitter), some of the ideas that form the basis for the (now thoroughly debunked[5]) Pizzagate conspiracy theory. This is a theory that went viral shortly after initial dissemination in 2016,[6] and which holds (among other things) that high-ranking US Democrat politicians, including Hillary Clinton, run a pedophile ring and perform Satanic rituals in a pizzeria called Comet Ping Pong.[7] Initially skeptical, Alex finds a seemingly credible detail that can be externally verified (e.g., that a politician's laptop was indeed seized by the FBI). This lends a sliver of credibility to the group.[8] When Alex questions why law enforcement isn't acting if the theory is true, he is directed to an unreliable source claiming an investigation is underway.

Gradually, he is drawn into dedicated forums like the r/pizzagate subreddit, where a comprehensive counternarrative is provided that purports to explain away all contrary evidence. By the end of this process, Alex is now "fully inside" the echo chamber. He distrusts mainstream sources (discredit), consumes information only from pro-Pizzagate sites (doxastic exclusion), and interprets debunking articles from outlets such as *The New York Times* as further proof of the media's complicity (defeater reinforcement). At this point, his rational means of escape are seriously limited. From his perspective, trusting the chorus of voices in the chamber and distrusting outsiders is the most rational course of action. This trajectory mirrors patterns observed in studies of online radicalization, where individuals are drawn into extremist ideologies through a gradual process of social reinforcement and the construction of a coherent fiction with (if its premises are granted) explanatory power (Alfano et al. 2021).

A very different kind of attractor-based loop we may encounter in any epistemic community where information flows freely involves *rumor diffusion*.[9] Here is a guiding analogy: In the case of rumor diffusion, information sharers contribute to an attractor-based loop in such a way that the driving function we play is, in this case, more akin to (vis-à-vis the star example) a gravitational field than to the objects that are being pulled by it. To make this idea more concrete, consider a simple case of particle diffusion (figure 7.2).

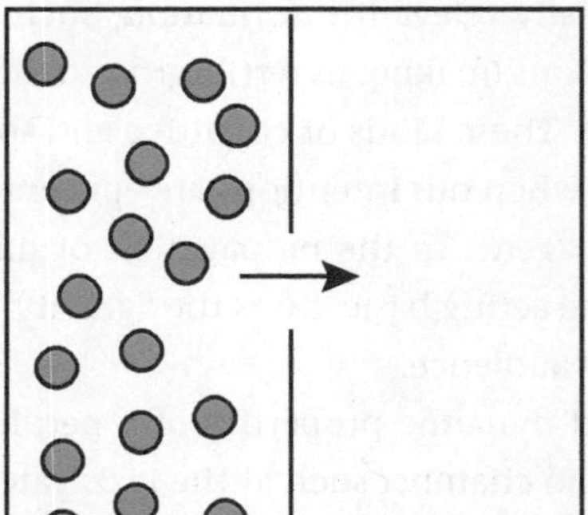

Figure 7.2
Particle diffusion (mass transport).

In simple "mass transport" diffusion, particles flow from a starting point of high concentration toward areas of low concentration. The movement of particles is associated with collisions with other particles (more frequent in high-density areas) which drive particles away from each other as they continue colliding (less frequently) over time, gradually reducing differences in concentration.

Here, the "attraction" is not of an agent *to* a group but rather of a piece of information *to* new uninitiated hosts. The agents themselves become the engine of the attractor's expansion. Empirical research has shown that false news and rumors (at least within certain demographics and platforms[10]) diffuse significantly faster, deeper, and more broadly than true news on social media platforms. A well-known large-scale study of Twitter by Vosoughi et al. (2018) found that falsehoods were 70 percent more likely to be retweeted than the truth, and that true stories took about six times as long to reach 1,500 people as false ones.

Several psychological mechanisms may explain this phenomenon. One hypothesis, offered by Sinan Aral (one of the coauthors of the Twitter study), is that false news is often more novel than true news, and humans are drawn to novelty (Bessi et al. 2015). Sharing novel information can also serve a social-signaling function whereby the sharer seems "in the know." Other research suggests that rumors conveying emotions such as anger and anticipation are particularly disposed to "go viral" (see, e.g., Pröllochs et al. 2021). Furthermore, our cognitive systems appear to prioritize and better recall socially relevant information, such as gossip and details about others' reputations, making us fertile ground for the spread of rumors (Mesoudi et al. 2006). These kinds of cognitive and social tendencies make us susceptible—even when our intentions are epistemically in the clear—to becoming unwitting agents in the propagation of misinformation and in such a way that we are acting basically as the "gravity" that pulls the rumor toward an ever-wider audience.

The structural and dynamic properties of a pernicious attractor, like a conspiracy theory echo chamber such as the Pizzagate case discussed previously, are often formally identical to those of a potentially beneficial one, such as an expert consensus group that dismisses fringe views. Take here Campbell's (2023) example of an antislavery consensus as a candidate virtuous echo chamber. Both (the good and the bad) operate by discrediting outsiders and reinforcing a core belief (regardless of the content of that belief). This means a self-governing agent looking not to be pulled into epistemically dangerous feedback loops cannot simply learn to identify and avoid a particular *structure* type. They must somehow (more on this to follow) make an evaluative judgment about the *content* of the attractor. This much motivates the need for a model of self-governance that is explicitly sensitive to the agent's own values.

7.2.2 Repeller-Driven Loops

Complementing attractors are *repellers*—that is, dynamics that push elements of a system *away* from each other (figure 7.3).

Magnetic force repels two things away from each other if they have the same magnetic charge. The same goes for electrostatic force. When positively charged objects get close to each other, they both push each other away from each other. In the social-epistemic landscape, repeller dynamics often manifest in a predictable and identifiable way: through polarization and the erosion of trust.

Group polarization is the tendency for deliberation among like-minded individuals to move them toward a more extreme position in the direction of their initial leanings. When two opposing groups deliberate, the typical result is bidirectional polarization: the groups are repelled further away from each other's views. Consider a simple example of polarization in the group case. A jury deliberating on a civil case might initially, before discussion, lean toward awarding a moderate amount of compensation to a plaintiff. After hours of discussion, they might end up agreeing on a very high figure—higher than any individual juror had thought appropriate beforehand. Or imagine a political discussion group of like-minded individuals (say, a group of environmental activists talking about climate policy). As they share ideas with each other, they might collectively arrive at proposals far more radical than any one member

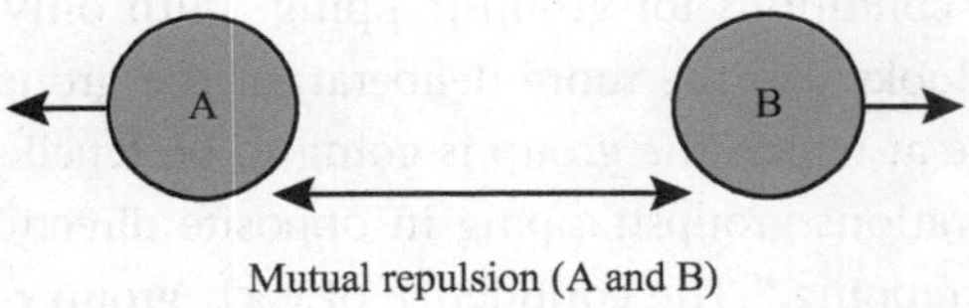

Figure 7.3
Mutual repulsion.

would have suggested alone. Cases like these, documented in both laboratory and real-world settings,[11] show the hallmark of group polarization (Broncano-Berrocal and Carter 2021, ch. 1–2). While several psychological mechanisms contribute to this, a particularly potent feedback loop described by Kenneth Boyd (2019) is a subset of group polarization cases called "groupstrapping."

Groupstrapping is a phenomenon that gets off the ground once two assumptions are made: (1) a group's justification for their belief is a (partial) function of the justification of the beliefs of its members[12]; and (2) groups can be testifiers.[13] If these assumptions are granted, space opens for the following evidence sharing and updating iterations to take shape:

> Consider a group G, and a member of that group, m. According to (2), groups can be sources of testimony (and, as we have seen, seeking out information from self-membership groups is not in itself illicit), so we can consider a case in which m seeks out information from G, and G testifies that p to m. On the basis of G's testimony, m then either comes to form a new belief that p, or else updates her degree of belief that p. Now, according to (1), G's warrant is a function of the warrant of its members. However, since m is a member of G, then since one of G's members becomes better positioned epistemically towards p, G itself appears to become more warranted in believing that p. It appears, then, that G has improved its epistemic position with regards to p by simply testifying to one of its own members, and without acquiring any new evidence. In turn, G appears to become a better source of testimony with regards to p. G and m, then, have engaged in a process of groupstrapping. (Boyd 2019, 67)

According to Boyd, then, confidence gained via bootstrapping from the group's view will continue ratcheting up as long as the process of consulting a self-membership group is iterated where individuals in the group continue updating on this kind of group testimony. With that in mind, it then looks like to get caught in a (repulsive) feedback loop, it would potentially be enough to engage in (a) deliberation within a group that bidirectionally polarizes, where (b) the deliberative process driving the bidirectional polarization meets the conditions for groupstrapping. With only those ingredients in play, it looks like the more deliberation the group undertakes, the higher the rate at which the group is going to be repelled away from each other via iterations groupstrapping in opposite directions. Call this "repulsive groupstrapping." The confidence of each group ratchets up in opposite directions, powered by this internal feedback loop, driving them further apart without any new external evidence being introduced. This cognitive process fuels the social phenomenon of affective polarization, which involves growing animosity, distrust, and dislike between partisan groups (Torcal et al. 2023)—and this emotional charge then serves to make the repulsion stronger and more socially corrosive.

7.2.3 Generative Loops

An arguably even more insidious repeller dynamic is generated by what we can call "unenlightened" star networks. A star network is a trust topology with a central hub and multiple nodes (trustees) who have trust relations with the hub but are (either by design or accidentally) prevented from communicating with each other (figure 7.4).[14]

Such networks are often created and then sustained intentionally by the hub to maintain power, as seen in financial frauds (e.g., Bernard Madoff's now well-documented Ponzi scheme), well-organized cults, and predator–victim relationships (Alfano 2016). The "repulsion" that features in this loop is between a node's belief state and the truth about the hub. The key mechanism is the exploitation of what Sanford Goldberg calls the "epistemic coverage inference": the rational assumption that "If that were true, I would have heard about it by now" (Goldberg 2011). In a healthy epistemic community, reasoning in accordance with epistemic coverage conditionals is a reliable heuristic for dismissing outlandish claims. (Imagine someone

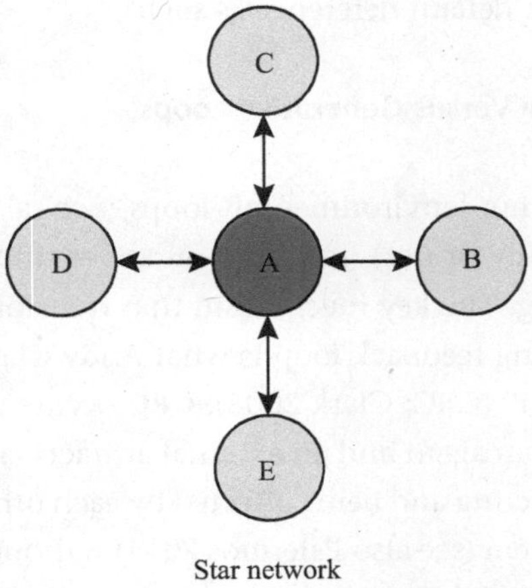

Figure 7.4
Star network.

tells you something sure to generate a lot of chatter, e.g., "There's been a new Royal wedding in Britain!" A natural dismissal to that assertion is: "If that were true, I would have heard about it!") In an unenlightened star network, however, the node occupiers suffer from higher-order ignorance. They are unaware that they are being systematically isolated from the testimony of other nodes. They do not know what they do not know. This makes the above kind of coverage inference systematically unreliable yet subjectively rational from the point of view of the node occupiers, given the evidence that will be available to them.

The feedback loop at play here operates as follows, using Madoff's star network as an example. An investor in Madoff's scheme hears no negative reports about what Bernie is up to. They infer from this absence of evidence that Madoff is trustworthy. This increased confidence strengthens the premise for their *next* coverage inference ("If he were a swindler, I *really* would have heard by now!"). Each iteration of this subjectively rational but unreliable inference ratchets up their confidence in the hub and repels them further from the truth, making them increasingly resistant to any external warning signs (Arvedlund 2009). This underscores the general moral: Deference and coverage-style heuristics are perfectly fine so long as we're in healthy networks. The loop-like repulsion from truth featuring in unenlightened star networks like Madoff's is grounded in the network architecture (selective isolation and discrediting), not by reliance on testimony or even sheer default deference as such.

7.3 Environmental Versus Generative Loops

Beyond encountering environmental loops, we also generate (again, whether intentionally or not) new ones ourselves through our patterns of cognitive offloading. The key mechanism that transforms discrete reliance into a self-reinforcing feedback loop is what Andy Clark calls "continuous reciprocal causation" (CRC; Clark 2008). CRC occurs when two systems—for instance, a human agent and an external artifact—are continuously and simultaneously affecting and being affected by each other, forming a single, tightly coupled system (see also Palermos 2014) A thought experiment contrasting "Otto" and "Telo" (a variation on a case by Rupert [2004]) illustrates this distinction well. Telo uses his phone book as a simple tool; the causal influence is just one-way: from the book to Telo. By contrast, Otto (the hero in Clark and Chalmers's [1998] paper "The Extended Mind") who has Alzheimer's disease, relies on a notebook that he carries everywhere. He constantly writes new information in it and consults it to guide his actions. The notebook structures his behavior, which, in turn, leads him to make new entries in the notebook. Otto and his notebook exhibit CRC. They have formed a coupled system, even though Telo and the phone book do not. (By contrast, discrete offloading *without* continuous reciprocal coupling [Telo-style use] is dependence without a loop; CRC marks the step where dependence becomes loop forming.)

Notice, though, that this generative loop is epistemically beneficial, even essential, for Otto (assuming, as per the case, his biological memory is failing). However, it also creates a deep dependency. The more Otto relies on the notebook, the more his biological memory may atrophy (or, to make the point, replace Otto with an ordinary cognizer who stops relying on biomemory), which, in turn, increases his reliance on the notebook. The loop is in this way *self-reinforcing*, and in a way that demonstrates that even loops we create for our own benefit can reduce our optionality and come with trade-offs in such a way as to make it ultimately more costly to exit the loop were we to choose to do so.

7.4 The Dilemma and Risk Opacity

The existence of this complex, loop-ridden landscape—where feedback loops are common (though by no means ubiquitous) in high-coupling, high-amplification settings—raises a tough question: how exactly are we supposed to secure the benefits of epistemic dependence while avoiding the traps, in the form of feedback loops, to which such dependence makes us vulnerable?

A first, intuitive reaction might be to view all feedback loops as epistemically detrimental simply on account of the fact that they restrict an agent's intellectual options. This view, though, is mistaken. As the case of Otto demonstrates, a generative loop can be epistemically beneficial, even in some cases *essential* to carrying on in everyday life and meeting our goals. More surprisingly, even environmental loops like echo chambers can be beneficial. Here again, recall Campbell's (2023) point that certain echo chambers can protect valuable and hard-won consensus from being undermined by frivolous or malicious challenges. (In his case, the thought was that a social-epistemic environment that exerts strong normative pressure to dismiss proslavery arguments out of hand functions as a beneficial echo chamber. It protects a crucial moral and epistemic consensus by discrediting and refusing to engage with a harmful position.) The goal for an (suitably) autonomous agent, therefore, cannot be to just go "scorched earth" and act (or omit actions) so as to avoid all loops simpliciter. It must be to gain capacities to discern—as well as we can, which won't be an infallible capacity—which loops to enter, which to avoid, and which to create.

This brings us to the dilemma we briefly teased in the opening section: Radical *independence* is (as the reader will have seen us argue for in various ways by now, including in chapter 5) epistemically perilous. Yet, as this chapter suggests, uncritical dependence leaves us vulnerable to being trapped in pernicious loops. So, the solution must lie in some alternative.

The positive thesis of this chapter is that the species of autonomy needed to navigate the loop-ridden landscape is a kind of capacity for *adaptive control*, which is itself grounded in the cognitive achievement of practical understanding.

7.5 Adaptive Control and Practical Understanding

One of the most common ways dependence turns pernicious is when it develops the features of a feedback loop without the agent recognizing it as such until it is too late. A quick diagnostic can help clarify what is at stake here. A dependency begins to risk becoming loop forming when (i) visibility is asymmetrically curated (e.g., via platform algorithms, star-network-style isolation), (ii) credibility is updated by the very signals the dependency generates (such as engagement, follower counts), (iii) countersources are preemptively discredited, and (iv) exit imposes salient social or practical costs. (Dependencies lacking these features, or at least most of them, typically remain corrigible and non-looping.)

This brings us to the primary practical difficulty facing a navigating agent: *risk opacity*.

(Ex ante) Risk opacity: A given epistemic loop is *ex ante risk-opaque* for an inquirer, *I*, to the extent that *I* is positioned to competently judge whether participating in the loop is in *I*'s epistemic self-interest (or not) only after participating in the loop.

Alex, in the Pizzagate example, could not have known ex ante that he was stepping into a pernicious environment. For him, the loop was ex ante risk opaque (hereafter, risk opaque). This problem is a special case of a more general problem addressed in engineering and control theory: the problem of *adaptive control*. Adaptive control systems are designed—at a high level of generality—to maintain a stable trajectory toward a desired goal, even when the environment is uncertain, nonlinear, and changing. At a conceptual level, these systems operate via three core strategies:

Parameter estimation: Continuously monitoring the environment and updating one's model of it.

Control law adjustment: Modifying one's own behavior and strategy in response to changes in the environment.

Feedback integration: Continually revising both estimations and adjustments based on feedback from the system's performance.

"Risk opacity," then, marks precisely those situations in which (when opacity is high) adaptive control is challenging, and discriminating good from bad loops prior to entering into their pull is just not feasible at the time. The salient question is whether one's *control competences* are such that, if the loop begins to misalign, inquiry remains resilient—via off-ramps, friction for escalation, cross-checks, and so forth—despite the lack of contemporaneous discriminative cues.

To see how an epistemic agent can implement adaptive control, it is necessary to articulate the cognitive requirements of autonomy. Autonomous action, whatever else it requires, needs more than mere intentional action. Drawing on the work of Reza Hadisi (2023), a crucial distinction can be made between practical knowledge and, more demandingly, practical *understanding*. Practical knowledge is, following Anscombe, the knowledge of the means one is implementing in doing what one does intentionally—that is, it is knowing that one is doing action A *in order to* achieve goal G. A core idea in recent action theory[15] is that the link between practical knowledge and intentional action is a tight one, especially in recent action theory. Consider that, for example, Frankfurt's unwilling addict acts with practical knowledge when she lights a cigarette in order to get a nicotine fix. She intentionally lights the cigarette. Does she autonomously light it? Perhaps not, after all, she has a higher-order desire to quit. In this way, while acting with practical knowledge of what one is doing is (more or less) good enough for acting intentionally (see here Pavese 2022), it is clearly not good enough for acting autonomously—and in this way, it would be a mistake to equate intentional action with what's needed for autonomous action.

What Hadisi terms "practical understanding" is a comparatively more demanding cognitive achievement than mere practical knowledge. It involves, following Hadisi, a veridical, dynamic grasp of the network of relations among one's actions, intentions, and values. In this way, practical understanding of what one is doing involves not just knowing what one is doing and why (this is something ensured by just having practical knowledge in action) but also possessing a kind of dispositional responsiveness to how one's actions cohere (or fail to cohere) with one's endorsed values over time. In this chapter, we'll use this term (practical understanding) in an epistemologically *externalist* sense, which we take to be recoverable from how Hadisi is using it in the purely action-theoretic case. By "practical understanding" of the sort that matters in intentional inquiry, we mean a competence for *adaptive control* over one's epistemic dependencies—a reliability-involving capacity to manage uptake, preserve effective exit options, and dampen misaligned reinforcement as conditions vary. Accordingly, practical understanding (so construed) is not going to reduce to or even require a perspective-bound knack for diagnosing one's situation across all good and bad cases. We should think of it rather as a modal feature of one's control profile across nearby possibilities when one is inquiring and which is realized in policies and abilities that would keep inquiry resilient *were reinforcement to begin drifting*.

Fully autonomous inquiry (inquiry that is not merely negatively autonomous in that it's sufficiently independent of assistance), as the thought goes, requires not just practical knowledge of what one's doing, in inquiring, but practical understanding of what one's doing. An agent can inquire intentionally (with practical knowledge) while failing to inquire autonomously because their actions (e.g., the sources they trust, the communities they join) have become misaligned with their deeper epistemic values (e.g., truth, accuracy, open-mindedness), and they lack the responsiveness to detect and correct this drift. On the present view, what matters (vis-à-vis maintaining autonomy in inquiry in loopy environments) is the reliability and availability of these control competences across nearby conditions—not a moment-to-moment ability to *tell* whether one currently occupies a benign or a malign loop.

The thesis we want to explore now is that the dispositional responsiveness inherent in practical understanding is precisely the cognitive faculty that enables adaptive control in the epistemic domain. An agent who possesses practical understanding to a sufficient degree (see Kelp 2015) is constitutively concerned with maintaining coherence between their actions, intentions, and epistemic values (including, e.g., values in intellectual ethics, viz., whether zetetically one values terminating in mere deferential knowledge, in firsthand knowledge, etc.). This concern creates an attentional and motivational profile that makes them *sensitive to signals that this particular type of coherence is breaking down* when it is. When such signals are then detected, the dispositional nature of practical understanding (for those who possess a sufficient degree of it for adaptive control[16]) motivates "course correction"—adjusting one's inquiry strategy. This process of monitoring for value misalignment and adjusting behavior just *is* an exercise of adaptive control.

So, what would such signals be exactly? Empirical and social science offers us a menu of them. These include the following (in no particular order, and with no claim to exhaustivity): (i) lack of diverse perspectives (Sunstein 2017); (ii) demonization of outsiders (Iyengar et al. 2012); (iii) insulation from criticism (Benkler et al. 2018); (iv) morphing of language (making it difficult for outsiders to penetrate; Nguyen 2020); (v) extreme confidence (Barberá et al. 2015); (vi) conspiracy thinking (Douglas et al. 2017); (vii) stifling of internal dissent (including through sanctions or threat of sanctions including social penalties; Stroud 2010; Leviston et al. 2013); and (viii) fixation on particular narratives (Garrett 2009; Katsafanas 2009). The presence of a cluster of these signals serves as feedback that the agent's current inquiry path may be leading them away from their epistemic values. A key intellectual virtue that facilitates the recognition of these signals and the willingness to act on them is intellectual humility (Whitcomb et al. 2016). This virtue involves an awareness of one's own fallibility and cognitive limitations, which fosters openness to corrective evidence and a willingness to question one's own epistemic situation.

Let's consider now an objection. In a *good* case (like the antislavery echo chamber), many of the warning signals from the cluster described in (i)–(vii) might well be absent; but such signals will *also* be absent ex hypothesi in the case of an evidence-resistant echo chamber that is ex post risk opaque. So, does this mean that Alex therefore would lack practical understanding (and by extension the kind of autonomy that implicates it) in the *good* case, and simply on account of the fact that he can't tell the difference?

Such thinking would rely on what is a highly contentious variant of the KK principle—a kind of "KU" principle—to the effect that if one understands something, X, then *one knows that one understands X*. If practical understanding of the sort of interest here (and which matters as a cognitive component for epistemic autonomy) is taken to obey such a "KU" principle, then indeed one would lack practical understanding in the good case if one can't (as one won't be able to) tell the good case from the bad case (e.g., when an evidence-resistant echo chamber is ex post risk opaque, such that no relevant signals are present).[17]

7.6 Concluding Remarks

The more we embrace epistemic dependence on others and on our cognitive scaffolding, the higher the premium becomes on our capacity to navigate the complex thickets of feedback loops that such dependence has a (defeasible) propensity to create. This chapter has aimed to chart out some of the key contours of this social-epistemic landscape by identifying three primary loop structures—(i) attractor driven, (ii) repeller driven, and (iii) generative—that can either support or subvert our epistemic aims.

We've suggested that it is a mistake to respond to this challenge by retreating to the flawed ideal of radical intellectual independence, an ideal we've chipped away at in one way or another in each of the chapters in this book. Such an ideal is, in short, not only impractical but epistemically disastrous. It offers (as we've registered in different ways) shelter to contrarians and conspiracy theorists. The form of epistemic autonomy worth valuing in the face of this particular challenge must be compatible with deep and extensive dependence. It requires a capacity to manage this dependence in a way that allows us to autonomously enter or generate loops that we recognize *do* align with our epistemic interests, while skillfully steering clear (to the extent we're capable) of those that would cut us off from them.

The position reached here is that this capacity best understood as a form of adaptive control, a concept borrowed from the study of dynamical systems. The cognitive engine of this adaptive control is, on the view we've laid out, not practical knowledge but rather practical understanding—a dispositional grasp of the dynamic relationship between our actions, intentions, and, crucially, our epistemic values. By cultivating this form of understanding, along with ancillary virtues such as intellectual humility, we, at the same time, help equip ourselves to better monitor our epistemic conduct, detect signs of value misalignment, and course correct as needed.

Notes

Chapter 1

1. For a more detailed discussion and development of this idea, see, e.g., Christman (1989) and Colburn (2010); for an overview, see Christman (2020).

2. There is, of course, a lively debate over what it takes for a given attitude or thought to count as one that is the agent's in the way required for autonomy. One influential approach asks whether a given attitude is such that one would, e.g., endorse it on reflection (Frankfurt 1971; Dworkin 1981); another is whether it was the product of compulsion (Mele 2001) or of other autonomy-subverting causes.

3. QAnon conspiracies, and their epistemic features, are taken up in more detail in our discussion of echo chambers in chapter 6.

4. Despite the narrative that conspiracy theory beliefs are increasing, research in the politics and psychology of conspiracy theories (Uscinski 2018) casts doubt on whether conspiracy theorizing has actually increased over the past century. As Uscinski notes, at nearly every period in US political history, that period had been proclaimed the golden age of conspiracy theories, albeit with no discernible evidence that genuine belief in conspiracy theories was actually rising at that time.

5. Epistemologists have been markedly more reluctant to accept this view than have social scientists. Rini (2021), de Ridder (2024), van Doorn (2023), and K. R. Harris (2022) have each argued that misinformation may threaten the informational environment and our warrant for our beliefs, even when it does not deceive. Simion (2024a) and Ichikawa (2024) dissent from the mainstream view in a different way: They still see the harm as primarily doxastic. For them, the harm of misinformation consists in its propensity to *undermine* belief. For Simion, "disinformation" *is* "ignorance generating content" (2024a, 6); for Ichikawa, the suspension of judgment is a major harm of misinformation.

6. It's worth noting that although we will cite Mercier quite extensively in making the case that fewer people form beliefs as a direct result of exposure to

misinformation than is often thought, his view is more extreme than ours. On his view, the *form* beliefs take may be the result of exposure to misinformation, but their central content is not. For example, whereas people may believe that vaccines cause autism as a result of exposure to misinformation, it is not such misinformation but rather generalized distrust and a variety of psychological dispositions that explain why they are suspicious of vaccines. They would very likely reject vaccination even in the absence of misinformation (Altay and Mercier, In press). Altay and Mercier may be right about that specific case—we are agnostic on that question—but we think that climate change skepticism is explained very largely by misinformation. Ironically, the view that misinformation is a more serious problem than climate change—a view that Altay and Mercier themselves mention—arises from misinformation.

7. For a detailed discussion on how epistemic vigilance operates not only personally but also sub-personally, see Rupert and Carter (forthcoming, chap. 2).

8. The growth of cities and the easy mobility characteristic of modernity have necessitated the development of institutional mechanisms for monitoring reputations. Vernon (2014) describes the development of the mechanisms that enabled trust and trade in Victorian Britain. Some of these same mechanisms underwrite *epistemic* trust in societies of strangers.

9. In a series of papers, Eric Mandelbaum and colleagues have argued that we reflexively believe every proposition we entertain, even when we already have conflicting beliefs (Mandelbaum 2014; Bendaña and Mandelbaum 2021; Porot and Mandelbaum 2021). We think the empirical foundations this view is built on are shaky. Even if they're right, however, the "beliefs" they identify have a different functional role from the beliefs we have in mind: They guide only a narrow slice of automatic behavior. We're using the term "belief" to pick out what Van Leeuwen (2023) calls "factual belief," where a factual belief governs inference and behavior whenever the agent is concerned with getting things right. We can use terminology however we like, of course, but factual beliefs play the central role in most behavior. Mandelbaum's belief fragments might underlie implicit biases, but implicit biases are poor predictors of behavior (Oswald et al. 2013).

10. According to Sperber and colleagues, our capacity for coherence checking (see also Graham 2023)—a form of filtering that targets the content rather than the source of communicated information—is one way in which this works. This capacity is explained by relevance theory: "The comprehension process itself involves the automatic activation of background information in the context of which the utterance may be interpreted as relevant. Here, the processing costs tend to be proportionate to the cognitive benefits derived" (Sperber et al. 2010, 374). This automatic activation serves a cost-effective function by leading us to monitor less closely assertions that appear less relevant to our background knowledge. On this view, "the same background information" used to assess relevance also provides a

cost-effective mechanism for coherence checking. In short, the cognitive processes involved in relevance seeking also function as a mechanism of coherence checking (Sperber et al. 2010).

11. Except, that is, when the model's actions seem to be a response to extraneous features of the situation. For example, if a model turns on a light switch by butting it with her head, children copy this technique, unless the model's hands are encumbered. In the former case, using her head seems to be part of the technique demonstrated; in the latter, it is explicable by the unavailability of her hands (Gergely et al. 2002).

12. Of course, in some cases where novices observe expert–expert disagreement in a given domain, the novice might simply get lucky and plump for an expert who has gotten things right; this trust is, in a minimal sense, well placed (in the minimal sense that a lucky belief might be true), but it remains incompetent, even if well placed in this minimal sense. For discussion here, see Carter (2024a).

13. In May 2025, the US Health Secretary Robert F. Kennedy Jr.'s office released a report under the banner of "Make America Healthy Again" (MAHA), which Kennedy described as a gold standard of scientific reporting. However, fact checkers noted that several of the studies relied on in the report, including ones that raised doubts about childhood vaccines, were backed by "phantom sources," citations that referred to nonexistent papers. This kind of error is without doubt a reason to downgrade the reliability of the report, but how is a novice to tell whether they should regard the errors as flaws in an otherwise reliable report or as fatal to its credibility? Trump's press secretary dismissed worries about the phantom citations as "formatting issues" (Jacobs 2025), suggesting that they were regrettable but not serious. Those who think the mistakes are *obviously* fatal to the report's credibility might do well to reflect on the controversy over the fourth Intergovernmental Panel on Climate Change (IPCC) report's claim that the Himalayan glaciers were likely to disappear by 2035, centuries earlier than the data suggested (Schmidt 2010). That claim was sourced to a World Wide Fund brochure, which, in turn, relied on an interview in *New Scientist*, which, in turn, cited a report that did not even attempt to estimate a date for the disappearance of the glaciers. We think the mistake, although highly embarrassing, does not cast doubt on the IPCC's report in any significant way. But it's not obvious why that should be so while at the same time the MAHA report should be dismissed due to its errors.

14. Not only is it true that values play a role in the assessment of scientific positions, we also think it is completely appropriate that this is the case.

15. We doubt, that is, that it is possible to show, on non-question-begging grounds, that the world we perceive is not a coherent deception of the sort Descartes imagined or (in more up-to-date versions) a simulation. The "problem of skepticism" might be construed in other ways. For example, it might be construed as the problem of

showing that our words refer, no matter what, or the problem of showing that our lives can be meaningful, no matter what. So construed, the problem may indeed have a satisfactory solution (but see Helton [2024] for doubts about the success of such solutions).

16. Contemporary foundationalists in epistemology are accordingly much more attracted to (fallibilist) moderate foundationalism. For discussion, see Carter and Littlejohn (2021, chap. 1)

Chapter 2

1. It might be argued that a domain of inquiry that is not concerned with belief or knowledge does not qualify as epistemology. Even if we restrict epistemology in this sort of way, we can't neglect attention: Attention plays a pervasive role in belief formation and in knowledge (Saint-Croix 2025).

2. We revisit some of these themes, to a different purpose, in chapter 5.

3. It's not at all obvious to us that those people who, say, share posts about QAnon really see its sources as sharing their values. They may, for example, recognize that Elon Musk amplifies conspiracy theories (M. Cohen 2024) for his own ends—ends they do not share. Nevertheless, they might delight in his posts on X because they target the "woke." They may be more motivated to rile their opponents than to promote any positive agenda (indeed, they may believe, with some justice, that no one has a realistic proposal that will really help them).

4. It may nevertheless give rise to all sorts of indirect harms. Maarten van Doorn (2023) has identified a number of harms that misinformation may give rise to, independently of whether it is believed. Its prevalence may induce general skepticism about the veracity of the information one encounters, it may (thereby) make it difficult to develop the virtue of intellectual autonomy, and it may increase support for unacceptably paternalistic policies. Although these concerns are serious, we think they are tractable when they arise from implausible misinformation. Most people give little credence to such misinformation. It should therefore have little power to induce general skepticism or to interfere with the development of epistemic autonomy. More skeptical reporting of belief claims in the media, so that people are no longer tempted to believe in the gullibility of *others*, should significantly reduce the temptation of unacceptably paternalistic policies.

5. What's the relationship between salience and attention? Roughly, we think of salience as an extra-agential determinant of attention. Attention is goal oriented and often under personal-level control. Salience is a fact about the extra-agential environment (although, of course, it may well be *indirectly* under agential control), which strongly and often automatically influences what we attend to (see Dicey Jennings [2022] for discussion).

6. On this issue, see the discussion, in chapter 5, of presuppositional erotetic nudging.

7. In recent papers focused on counterspeech, Cepollaro et al. (2023, 4) and Lepoutre (2024, 573–574) have noted that harmful speech may sometimes be harmful due to its capacity to command attention. The harms both papers focus on are doxastic harms: Salience may increase the fluency with which we process information, which is alleged to make it more plausible and may serve to give the impression that it has more evidential support than is really the case. Although these claims may be true, when misinformation is truly bizarre, any such effects do not bring the claims close to the threshold for being taken seriously as candidates for belief. Both papers also recognize, however, that there may also be a non-doxastic harm brought about when speech becomes salient: Unbelieved misinformation may distract us from more pressing concerns.

8. That is, even if no one takes it seriously *as a candidate for belief*. The harms we're concerned with largely arise when a significant proportion of people *pretend* to take it seriously in this way. If a claim is widely enough seen as too implausible to be given credence and almost no one is willing to defend it, it might become a running joke. In that sort of case, its being attended to might be harmless.

9. It's important to distinguish between genuine obliviousness and the performative obliviousness of those who decide some cultural phenomenon is beneath them (as well as those who don't pretend to obliviousness but choose not to participate in a discussion). The distinctions mark the difference between choosing and being unable to engage in conversations.

10. People sometimes argue that patterns of attention reflect prejudices. For example, it is sometimes said that the attention Americans pay to human rights abuses in Israel—compared with, say, the relative neglect of human rights abuses in Saudi Arabia or Kazakhstan—reflects anti-Semitism. Patterns of attention may indeed reflect (Siegel 2017), or constitute (Munton 2023), prejudices. But attention may be self-sustaining. We attend to Israel, in important part, *because we have attended to it*. Having attended to it made it central to our web of political attitudes and gave us a context in which to understand it. Its having been attended to makes it salient for us in a way that tends to assure continued attention.

11. Brooks (2023) points to a closely related phenomenon: the fertile ground created for conspiracy theories when public epistemic authorities fail to respond or respond inappropriately to apparently anomalous data. Third-party expectations impose epistemic obligations on us, the failure to abide by which has predictable effects on the audience.

12. This was written prior to the UK Supreme Court judgment that found that biological sex determines gender for the purposes of the Equality Act. Keir Starmer took—or thought it expedient to pretend that he took—that decision to settle the

extension of "woman" in ordinary language (McKiernan and Zeffman 2025). That truly bizarre interpretation of the judgment no doubt reflects the success of the attention-grabbing strategy at changing the shape of the debate.

13. Further, it is easy to increase the salience of this sort of topic, but hard to decrease it, for reasons that are not well understood. Simpson (2013) suggests this sort of asymmetry might arise because identity-prejudicial remarks or insinuations introduce networks of associations, and that attempts to cancel or undercut associations tend to reinforce them.

14. That many people believe that *p* is genuine evidence in favor of *p*. But it's defeasible evidence. Their attending to *p* is also defeasible evidence that *p* is worth attending to, but even if we subsequently judge *that p is not worth attending to*, we may find ourselves having to attend to *p* nevertheless. Attention may confer significance on something we would be better off not attending to.

15. Jenny Saul's recent book is, among many other things, a depressing taxonomy of the techniques Trump and those who attempt to emulate him use to this end (Saul 2024).

Chapter 3

1. In the wake of Bernard Williams's (1973) famous article, a rich literature on the degree and the ways in which belief may be voluntary has developed. As that literature has often emphasized, it's certainly true that we can voluntarily influence our beliefs, and that we can even engage in voluntary actions with the aim of making ourselves believe some proposition we take at the outset to be false. Belief remains involuntary in a central sense, however: One cannot believe that *p* if one also concurrently believes that there is or is likely to be a preponderance of evidence that conflicts with *p*. The person who successfully brings themselves to believe that *p* must also believe that the evidence genuinely supports that belief at the end of their program of belief inculcation, either by forgetting that they deliberately brought their belief about or by taking themselves to have luckily induced in themselves a belief that is independently supported by good evidence.

2. Most psychologists maintain that beliefs can be motivated. One way of construing motivated beliefs has them formed—involuntarily—in response to non-evidential considerations, such as a need to shore up one's sense of self-worth. If this is the case, then it's false that beliefs respond solely to evidence. But this is not the only defensible way of understanding motivated belief formation. Another possibility is that the motivation affects what evidence the person exposes herself to, what she recalls, and so on. It's consistent with the evidence that motivated belief is not an exception to the rule that belief formation responds to evidence alone, although it is also consistent with the evidence that belief formation responds at the margins to other forces (see Coppock [2023] for discussion).

3. Terminological note: Van Leeuwen thinks it is better to say that religious attitudes are not beliefs *of the same sort* as "factual beliefs" rather than saying they're not beliefs at all. We prefer to reserve the term "belief" for states with the functional roles of factual beliefs. We think this better comports with how "belief" is used in philosophy and cognitive science.

4. Van Leeuwen's criteria are, broadly, functionalist. Some philosophers have defended group beliefs on interpretivist grounds (Tollefsen [2002] has been influential). On this view, all there is to being a believer is to be interpretable from the intentional stance (Dennett 1989). On our view, interpretationism converges with functionalism because an entity that departs from the functional profile we associate with being a believer is better interpreted as guided by a different mental state (correlatively, we deny Backes's [2021] claim that intepretationists should not distinguish between group beliefs and group acceptances).

5. Schwengerer (2022) argues that we need not account for group lies at all because cases in which we are tempted to see a group as lying involve some other normative failing, and we can condemn the group on that basis. He thinks that these other normative failings account for our disposition to think the group is lying. Although the failings he points to are genuine, we do not think they are serious enough to account for the opprobrium rightly leveled at group agents in the kinds of cases Lackey considers. More pointedly still, the success of his rebuttal would not go far in establishing that joint acceptances are really beliefs: The underlying problem is not accounting for moral opprobrium, but rather accounting for the centrality of evidence in belief formation and maintenance.

6. Here, of course, we have in mind the view that part of what it is for something to count as a belief at all is that it is governed by a truth norm, such that it is correct iff true (Shah 2003; Shah and Velleman 2005).

7. Lackey's view is not pure summativism in that it places a constraint on when members' beliefs entail or constitute a group's belief: It must be the case that "adding together the bases of their beliefs that *p* yields a belief set that is not substantively incoherent" (2021, 49). We suspect this latter condition is better thought of as a condition on whether a group has a *justified* belief in virtue of its members believing that *p* rather than a condition on belief possession at all.

8. A reviewer for the MIT Press notes an interesting consideration against summativism. There are cases in which a group believes that *p and* the individuals who compose that group, or a significant proportion of them, believe that *p*. But summativism seems to get the direction of causation wrong: In these cases, the individual believes that *p because the group does* and not vice versa. Scientific research yields many such cases: A lab member may be agnostic whether *p* or even hostile to it until the paper is drafted; at that point, they accept that *p* on the basis of their own group's testimony. Our distributed model allows for such cases.

9. It's not.

10. There are many cases in which one and the same individual belongs to two different research groups, which assert conflicting claims. On occasion, there are even three-way conflicts: An individual *A* might belong to group *g1*, to which is appropriately attributed the belief that *p*, *g2*, which believes that *q*, while at the same time privately believing that *r*, where *p*, *q*, and *r* are incompatible propositions.

11. Note that this sensitivity to evidence must be *direct*. Jury cases, of the sort that motivate joint acceptance cases, do not satisfy our criteria because, in them, sensitivity to evidence is at best indirect. The rules governing the admissibility of evidence in a court of law might be justified on epistemic grounds, but they're aimed at classes of evidence, not particular instances of it: Evidence is inadmissible when allowing evidence of this *sort* leads to unsafe verdicts. When evidence that is known to be reliable belongs to such a class, it is excluded, despite its actual reliability. In scientific research, evidence is not excluded on this sort of basis: It is actual evidential value and not the kind of evidence that it is that matters.

12. It's also worth noting that Bird's account cannot satisfy Lackey's (2021) group lies and group bullshit desiderata for reasons directly related to its failure to satisfy the poise condition. A group to which knowledge is available only in the weak sense Bird takes to be necessary cannot intentionally mislead; such behavior requires that the relevant content be at least dispositionally available to it.

13. We do suspect that there are fewer bizarre beliefs than most philosophers think; fewer, most especially, beliefs that count as bizarre by the standards of the person's own culture. Monothematic delusions, such as Capgras (the delusion that a person close to an individual has been replaced by an impostor) or mirror misidentification (the delusion that a person's reflection is a different individual), are *defined* by the *Diagnostic and Statistical Manual of Mental Disorders* as "false beliefs" that are not "ordinarily accepted by other members of the person's culture or subculture." But delusions typically fail to satisfy our conditions for belief (see Bortolotti [2022] for discussion; Bortolotti herself defends doxasticism).

14. Bird (2010) cites cases of this sort to a different end: to show that individual members believing that *p* is not sufficient for a group belief that *p*.

15. If Levy (2024b) is right, however, the best explanation for such failures of inference and behavior is, in many cases, consistent with the individual-level state being a belief. In these sorts of cases, the attitude of group members will be a genuine belief with a sparse content.

16. Peer review in science is typically single anonymous: Reviewers know the identity of authors but not vice versa.

17. It's common to explain herding as an attempt by pollsters to avoid the reputational damage of being the lone outsider making a wrong call (P. Whiteley 2022). We by no means reject that explanation entirely, but it should be obvious to anyone

who's kept up with the literature on the epistemic significance of disagreement that there are rational pressures in the direction of conformity. It's worth adding that if there are risks in being an outlier, there are also large potential benefits: The outlier who gets it right experiences a big boost to their reputation. We would bet that better established polling firms would err on the side of minimizing risk, but newcomers should be prepared to take the gamble. Herding across established and newer polling firms probably largely reflects rational pressures.

18. An exception concerns subcultures: A subgroup of a bigger group might conceivably be oppressively socialized by the larger society (independent of how its individual members are socialized).

19. Tuomela's (1992) account is slightly different: On his view, the acceptance of the belief by group members presupposes the acceptance by others rather than being conditional on it.

20. On Bird's distributed account, group knowledge requires nothing more than that the knowledge that *p* be available to the group "in principle." The fact that knowledge need not be easily available to the group or to the individuals who compose it leaves room for potential manipulation: Agents might deliberately make it harder to access the knowledge without removing the possibility of accessing it altogether. Intuitively, such manipulation might be a threat to the autonomy of group knowledge, although we confess we're unsure what to say about such cases.

21. The historicism just sketched is *negative*, requiring that agents or groups lack an autonomy-bypassing past. There are also positive historical accounts on the market, which require that agents *possess* an appropriate past (see Haji [2008] and McKenna [2012] for discussion). We think an application of this sort of historicism to groups is implausible. We take no stand on negative versus positive historicisms with regard to individuals, but we find the requirement that a group possess a past to be unmotivated as a condition for autonomy.

Chapter 4

1. Millgram is especially concerned with risks arising from the difficulties that different specialties face in conveying their results and theories across disciplinary boundaries in a way that adequately captures the caveats and qualifications that are the currency within each, such that the inputs each receives from the others are distorted. We do not share his sense of the size of the risk. We think that, in many areas, there are sufficient bridging experts available to play the role of ensuring that qualifications are appropriately taken into account. That's not to say that distorting simplifications do not occur—they do. It is to say that science has been able to avoid or mitigate the risks in those areas that are sufficiently mature.

2. LLMs are sometimes dismissed on the grounds that they are merely autocorrect on steroids (or "stochastic parrots"; Bender et al. 2021). They merely predict the

right pattern of words rather than actually engage in intelligent activity (let alone thought). There's a powerful response to this line of argument: On an increasingly influential view in cognitive science, *brains* work in a similar way. On this view, brains are composed of a large number of hierarchically organized mechanisms, each of which has as its primary activity the prediction of input (perceptual or from other mechanisms) and the update of its expectations on the basis of the difference between actual and predicted input. Brains, too, might count as stochastic parrots (see Friston [2009, 2010] for enormously influential defenses of this sort of view, and Hohwy [2013] and Clark [2016] for philosophical development).

3. We do not accept Marušić and White's suggestion that deliberating together must take the form of reasoning from the same body of evidence (2023, 218) or that consensus must be its aim (221). Inquiry together does require agreeing on some things, such as what counts as evidence, but it need not aim at consensus, and it tolerates a great deal of dissent about method and even some dissent about evidence. There has recently been a rise in what are called "adversarial collaborations" in science. They aim at identifying common ground and in making disputes sharper, not at reaching consensus (Rakow 2022).

4. There are also cases of peer disagreement in which we might respond by relegating *ourselves* from the status of epistemic peer, despite having, what seems to us, a capacity to encompass our considerations in a single mental act or state. Consider, for example, cases in which one suspect one might be impaired by a drug or suffering from hypoxia (Elga 2008). In such cases, we are alienated from our own judgments and the considerations that support them. But that's the appropriate attitude to take in such cases. We must take the objective attitude toward ourselves when we have good grounds for suspecting that we temporarily lack the capacity to be fully rational beings.

5. Repeated disagreement with a peer is much more of a threat to our status as rational agents than one-off disagreement. One-off episodes are easily explained away: I wasn't at my best that day, I slipped, I was tired, and so on.

6. LLMs that are at the AGI or ASI level might also scaffold the development of epistemic autonomy because—as we've seen—disagreement with an epistemic superior is not experienced as threatening to our status as rational agents.

7. It's worth noting that LLMs can *also* threaten autonomy precisely by undermining the capacity to develop the skills to disagree. They are often sycophantic toward their users (Malmqvist 2024). This sycophancy stunts the development of a variety of skills, including those specific to whatever domain of knowledge the user might be exploring, by producing a premature sense of mastery. Ironically, sycophancy might threaten autonomy by producing a sense that the person is self-sufficient.

8. One might deny that LLMs produce testimony on the basis that their outputs are literally meaningless. We're agnostic on that claim. On our view, if LLMs cannot produce testimony, they nevertheless are a source of what Mallory (2023)

calls "quasi-testimony." Their outputs convey information to appropriately placed interpreters.

9. The singularity, in this context, is a moment of rapid acceleration in intelligence at which self-designing and self-modifying AIs surpass us dramatically. At that point, AI development becomes uncontrollable. Some AI researchers fear that, at this point (which many predict is no more than decades away, and some suggest might even be imminent), our creations will so surpass us that they will not feel any compunction in sweeping us aside. We take no stand on the likelihood or imminence of the singularity.

10. We adopt Sean Michaels's (2023) evocative description of using an LLM as a cognitive tool, but the "slot machine" language doesn't capture how creative usage typically is. Michaels himself fed lines of poetry into ChatGPT in order to generate poems for his novel about a poet collaborating with an AI. The slot machine description is justified by the fact that he repeated the process multiple times, selecting among the lines the algorithm generated. Typically, users will engage more creatively with the AI, carefully crafting prompts and often refining them in response to the output—quite often engaging in what appears to be dialogue with the LLM.

11. Duede (2022) argues that our justification for accepting the outputs of deep learning algorithms cannot be understood on the model of our justification for accepting either the outputs of tools or the outputs of experts. Our justification for accepting the outputs of computational tools depends on our grasp of the theories upon which they rely, but deep learning algorithms are opaque. Our justification for accepting the claims of (expert) agents depends on our in-principle access to the first-order reasons upon which they rely. Duede's argument provides support for our agential ambiguity claim; LLMs are neither quite tools nor quite agents.

12. In the wake of Fricker's (2007) enormously influential work on epistemic injustice, social epistemology has emphasized that oppressed people usually have better insight into their circumstances than they are given credit for. The epistemic deficit they suffer from is not theirs; it is the deficit of *others*, who can't or won't hear their valid complaints (see also Mills [2007] for how ignorance might be adaptively produced on the part of dominating groups). Although we entirely accept that members of oppressed groups suffer from a pervasive problem of testimonial injustice we believe that the phenomena of adaptive preference formation and oppressive socialization are equally real. See Brancazio and Levy (2024) for discussion.

13. In some cases, I might even be unable to identify components of the work package that are largely due to me. The inability to take responsibility for the overall output is independent of whether we can take responsibility for components of it, although the problem is worsened when we cannot do the latter.

14. The obvious ways for a paper to bear an agent's mark is for that paper to contain sentences written by the agent or sentences expressing ideas that were originally

written by that agent. But there are other ways for a paper to bear the mark of an agent. Perhaps the agent wrote a paragraph that was subsequently deleted but which was instrumental in bringing about contributions by other agents. In some cases, that sort of contribution might be significant enough for the agent to deserve credit for the final product.

15. These reflections might give us grounds for responding to Quill Kukla's (2012) worry that given how deeply distributed authorship in science is, it's difficult to see why ghostwriting a scientific article (a sadly real practice) is problematic. After all, the ostensible author—who is always a genuine scientific authority—might have every reason to be confident in the findings they put their name to, and needn't stand in a very different relationship to those findings than someone who was actually in a position ethically to claim authorship. But there's an important difference between the ostensible author and, say, the lab head who contributed only a small amount to the overall product. In the second case alone, the product reflects an epistemically autonomous contribution of the author; they played a role in its production. It's a sufficient condition of authorship that the person played such a role, even if they cannot claim credit for the product itself. A difficult case remains: What about the ghostwriter who was not involved in the research reported but whose intellectually autonomous activity is reflected in it nevertheless (because they have helped to develop crucial intellectual apparatus the work builds on, for example)? We think that such ghostwriting might be acceptable, although we do not have a firm view here.

Chapter 5

1. As Reid had put it, for such an inquirer, "no sentence uttered in the speech would be believed, until they were examined and tested by reason, and most men would be unable to find reasons to believe in a hundredth of what they are told" ([1764] 1997).

2. To use an even more high-tech example, to determine the number of logs in a truck bed, you could just count them visually or use an augmented reality app such as CountThings from Dynamic Ventures to scan and count them for you and spit out an answer (e.g., 34 logs). See https://www.countthings.com.

3. For discussion on this point, see, e.g., Pritchard (2009, 2014).

4. For some related discussion, which goes beyond our more targeted cases in this chapter, see Levy and Alfano (2019).

5. For simplicity, we can assume the proof is done a priori, with no aids at all.

6. Denying this would be disastrous. We're assuming, of course, that the textbook is a normal, authoritative textbook and that there are no undefeated defeaters for mathematician B. So, regardless of whether one construes the epistemic situation of

receiving the author's testimony on reductionist or non-reductionist lines, any plausible (non-skeptical) epistemology of testimony generates clearly the same result that by consulting the book you can get not only justified belief but also knowledge.

7. See especially Sosa (2021; cf. Carter 2023). Sosa's 2021 version is an extension of his earlier performance-theoretic virtue epistemology developed in Sosa (2007, 2009, 2010, 2015).

8. The most recent versions of this picture, especially in the latter chapters of Sosa (2021), introduce even higher grades of knowledge, such as secure knowledge full well. The most important innovation of this new theory, which goes beyond what we'll discuss here, is his introduction of default assumptions in epistemology, which are related closely to background conditions entailed by the obtaining of the conditions pertinent to the exercise of a competence.

9. Sosa thus takes it that an apt belief can be attained even when you simply trust a speaker in a good case. This is an important difference between Sosa and, on the other hand, Pritchard and Lackey, who deny that, in testimony cases, the cognitive success attained is sufficiently "because" of the exercise of the testimonial recipient's ability. It is worth registering that this dispute turns on how to unpack the "because." For Sosa, aptness requires "because" in the sense just of the belief's correctness manifesting complete competence manifestation. Greco, by contrast, requires causal-explanatory salience, which less plausibly will be met in simple testimony cases.

10. See https://mathstodon.xyz/@tao/111206761117553482.

11. The very idea that, in solving new problems, one is nonetheless standing on the shoulders of giants featured prominently in a widely covered incident following Grigory Perelman's celebrated proof, in 2003, of the Poincaré Conjecture, one of the Clay Mathematics Institute's "million dollar" problems. Perelman was offered the money but refused it, claiming that Richard Hamilton, who pioneered the Ricci Flow technique, which Perelman relied on, was at least equally deserving. For the story in more detail, see Nasar and Gruber (2006) and Gessen (2011).

12. Consider here North Korea's *Juche*, or "self-reliance" economics.

13. Also it's not clear this is physically possible, given limitations on how the brain is able to generate mathematical thinking. See Spelke and Tsivkin (2001).

14. See Stoljar and Mackenzie (2000).

15. For a recent defense of this idea, see Boult (2024).

16. The basic distinction matters here. Even in cases where independence is systematically useful (one might *want* to do a problem by themselves, and so independence is trivially needed to achieve that, or one might *need* independence from chatter to focus, etc.), it does not follow that exhibiting independence is to one's *epistemic*

merit in the way getting knowledge, understanding, explanatory insight, and so forth is.

17. We take this kind of scenario as a central focus in a later chapter on epistemic nudging and interrogative attitudes.

Chapter 6

1. For some overviews of the general problem, including in connection with the mis-/disinformation distinction, see Simion (2024a, 2024b), Simion and Kelp (2025), Del Vicario et al. (2016), K. P. K. Kumar and Geethakumari (2014), and Levy (2023).

2. For variations of this criticism, see, e.g., Moran (2021) and Bennett (2010).

3. Critics of the approach in political philosophy and various branches of applied ethics worry that nudging is *manipulative* (remember, there was no explicit argument here for the fruit) and so, in short, achieves the desired behavioral outcome at the expense of suitably respecting autonomy. The template manipulation argument goes like this. Manipulation undermines autonomy. What distinguishes manipulative from non-manipulative influence is that, in the former case, one influences another by *bypassing their capacity for reason*. Sufficient for such bypassing is that the influence either exploits their nonrational psychology or influences choices in a nonobvious way (Blumenthal-Barby and Burroughs 2012, 5). Influencing someone's choice via nudging seems to do both of these things (although this point is challenged in sections 6.5 and 6.6). Therefore, as the thought goes, nudging is manipulative and thus undermining of autonomy. For some representative discussions here, see, e.g., Bovens (2009) and Wilkinson (2013). Proponents take the manipulation worry to be overstated. For instance, in recent work, Levy (2019b) has pointed out that it's far from clear that nudging actually *bypasses* our reasoning processes in the first place. This point will be developed later as applicable to interrogative attitudes.

4. See Duncker and Lees (1945). For discussion of the functional fixedness bias in the context of epistemology, see Alfano (2012).

5. For a more recent review of literature on the psychology of the anchoring effect, see Furnham and Boo (2011).

6. We borrow this example, albeit from initially a different philosophical purpose in the literature on know-how, from Bengson and Moffett (2011).

7. As Grundmann asks, "What does it mean to say that the nudging of certain doxastic attitudes preserves the nudgee's rational control concerning these attitudes? The main point here is that when doxastic attitudes are triggered by nudges, this does not remove the nudgee's ability to revise these states in the light of further evidence. Thus, nudging does not isolate doxastic states from rational influences, as indoctrination or brainwashing would do" (2023, 211).

8. See chapter 2 for related discussion of hijacking in relation to attention.

9. It's worth noting that the account here looks compatible with a key point in Simion's (2024a) account of disinformation, on which disinformation is information disposed to generate ignorance. Choice architecture manipulation that results in erotetic nudges that lead to ignorance (imagine a case of an erotetic nudge that leads one to ask a question with a false presupposition rather than a sound question) looks as though whatever information features in the choice architecture tweak ought to be a wide form of disinformation. This looks right, and Simion's account predicts this result.

Chapter 7

1. For some relevant discussions here, see, e.g., Alfano et al. (2020), Clark (2012), Heersmink (2015), Wilson and Foglia (2017), Palermos (2020), Andrada (2019), and Menary (2012).

2. For more detailed discussion, see Beer (1995), Abraham et al. (1990), and Palermos (2014).

3. A terminological clarification: We are following Nguyen here insofar as he takes it that echo chambers have a structure "from which other relevant voices have been actively excluded and *discredited*" (2020, 141, italics added) and that echo chambers involve "excluding non-members through epistemic discrediting, while simultaneously amplifying members' epistemic credentials" (2020, 146). Now, that said, it's worth noting that echo chambers are often discussed in the literature in a way that incorporates a principle weaker than defeater reinforcement, which we can call "evidential preemption"—namely, the viewpoints of outsiders are evidentially preempted: They're assigned zero credibility. We are happy to include structured groups that embrace evidential preemption rather than defeater reinforcement under the broad banner of an echo chamber. For our purposes here, the former variety (which takes on defeater reinforcement) is especially epistemically dangerous (and pertinent from the perspective of feedback loops, specifically), and so they will be our focus.

4. Elzinga (2022) suggests a further component of echo chambers: that they feature what he calls "collaborative resistance." As far as we can tell, collaborative resistance in Elzinga's sense might be implied by the conjunction of the other three elements described here.

5. For an overview not only of how the conspiracy was debunked but also of how it plausibly gained initial traction despite its lack of credibility, see Bleakley (2023).

6. See https://www.nytimes.com/interactive/2016/12/10/business/media/pizzagate.html.

7. See Metaxas and Finn (2019).

8. As Cassam notes, this kind of tactic (mixing conspiracy with fact) is a typical one, as it "creates the impression that his theories are the product of reliable research into trustworthy sources" (2019, 20).

9. For discussions of rumor diffusion, see, e.g., Vosoughi et al. (2018), Cheng et al. (2013), and Takayasu et al. (2015).

10. Subsequent work following the Vosoughi et al. study cautions against treating this as a platform/demographic-independent generalization. The overall production and sharing of misinformation appears concentrated among certain demographics of users (Atlay et al. 2023), and reputational costs deter most people from sharing falsehoods. So, the breadth of the problem may (in these earlier studies) be overstated.

11. For discussion, see Broncano-Berrocal and Carter (2021), Singer et al. (2019), O'Connor and Weatherall (2018), Bail et al. (2018), De Cruz (2020), and Isenberg (1986). For discussion generally of polarization in epistemology, see Carter (forthcoming).

12. As a point of clarification: This assumption needn't be read as committed to the denial of the thesis that some groups have inflationary (i.e., non-summative) properties. Those who embrace non-summativist accounts of various group-level phenomena aren't committed to denying that there are at least some instances where a group-level property will be one it has in virtue of some aggregation of the properties of the group's members. So, for the assumption being made here in the context of groupstrapping, we need only assume that some group-level epistemic properties can be given a summativist reading. For discussion on the summativist/non-summativist distinction in group-level epistemology, see Broncano-Berrocal and Carter (2021) and chapter 3 of this book.

13. For a notable recent account of group testimony, see Lackey (2018).

14. For a helpful discussion, see Alfano (2016, 44–45).

15. In particular, this is given the recent turn in tide away from accepting the force of Davidson's (1970/2021) carbon-copier-style case against knowledge conditions on intentional action.

16. What, concretely, follows for agents navigating maximally *risk-opaque* environments? Unfortunately, very little. The recommendation here (in a "bad case") is modest but actionable, and follows from our conditions on adaptive control. When navigating risk opaque environments, our best policy for attempting adaptive control is to budget dependency and introduce friction once reliance escalates; install cross-cutting checkpoints that sample credible but non-cooperative sources; and pre-commit off-ramps that trigger when reinforcement starts to ratchet. These are control-level policies that give at least a chance to sustain (some degree of)

autonomy in our inquiries by building on our minimal level of practical understanding. In such circumstances, discriminating good from bad cases in real time is difficult. Where we cannot gain a sufficient degree of practical understanding to support a capacity for adaptive control, we are to that degree less autonomous.

17. But there are well-known reasons to resist iterative principles in epistemology such as KK (Williamson 2000) not least that they fall foul of margin-for-error-style sorties arguments. KK in an outlier thesis in epistemology (though cf. D. Greco 2014). Rather, the prevailing view is that knowledge isn't luminous (viz., the prevailing view is that it's not the case if we are in a knowledge state we thereby know we are). And this point transmits largely over to existing thinking about understanding. Even if we often know that we understand things, it's likewise contentious that we understand something *only if* we know that we do. Here is an example: I might, as it were, understand physics despite doubting that I do and so despite not knowing that I do. In such a case, I might understand physics perfectly well while entertaining misleading testimony to the effect that I have, say, been reading from a misprinted textbook. Our response to the envisioned objection can now be put as follows: The idea that one could have autonomy-relevant practical understanding in the good case only if one can tell the good case from the bad case is an idea that gets up and running only if one embraces a contentious kind of epistemological principle (KU), a principle that should be contentious for reasons that KK is contentious. We have no good reason to accept a KU principle (which implies that practical understanding is luminous) and, by extension, no good reason to doubt that one could have autonomy-relevant practical understanding in the good case (e.g., an echo chamber that aligns with one's epistemic interests, thereby lacking signals that it is an evidence-resistant echo chamber), even if one can't tell the good case from the bad case where such signals would be in principle undetectable ex hypothesi. The echo chamber example that's been guiding our discussion in this section, it should be noted, is meant to be illustrative of loops more broadly. If, given the risk opacity of various loops we might easily get pulled into, the implementation of adaptive control is beneficial in staying on course (to the extent feasible; on course *by our own lights*—viz., given what we value epistemically), then development of this underappreciated dimension of epistemic autonomy (practical understanding) affords us just this, even if in certain bad cases (those featuring ex post risk opacity) there is little we can do. Apart from echo chambers, the point applies in, e.g., say, unenlightened star networks (which will tend to seal us off from our real interests in most all cases), as well as generative loops, which often, but needn't always, promote them.

References

Aamodt, Michael, G., and Heather Custer 2006. "Who Can Best Catch a Liar?" *Forensic Examiner* 15 (1): 6–11.

Abraham, F. David, Ralph H. Abraham, and Christopher D. Shaw. 1990. *A Visual Introduction to Dynamical Systems Theory for Psychology*. Aerial Press.

Acerbi, Alberto. 2019. "Misinformation." In *Cultural Evolution in the Digital Age*. Oxford University Press. https://doi.org/10.1093/oso/9780198835943.003.0006.

Adams, Frederick, and Kenneth Aizawa. 2008. *The Bounds of Cognition*. Wiley-Blackwell.

Adams, Michael. 2012. *Slang: The People's Poetry*. Oxford University Press.

Adamson, R. E. 1952. "Functional Fixedness as Related to Problem Solving: A Repetition of Three Experiments." *Journal of Experimental Psychology* 44(4), 288.

Adekola, Josephine, Jamila Garba Audu, Thelma Okey-Adibe, and Aliyu Abubakar. 2022. "COVID-19 Vaccine Experience and Hesitancy Within African, Caribbean, and Black (ACB) Communities in Scotland." BEMIS. https://publichealthscotland.scot/repository/covid-19-vaccine-experience-and-hesitancy-within-african-caribbean-and-black-acb-communities-in-scotland/.

Adekola, Josephine, and Mona Simion. 2025. "The Infodemic, Epistemic Exclusion in Science Communication, and Distrust in Scientific Expertise." In *The Epistemology of Experts*, edited by Peter Brössel, Anna-Maria Asuntat Eder, and Thomas Grundmann. Routledge.

Alfano, Mark. 2012. "Expanding the Situationist Challenge to Responsibilist Virtue Epistemology." *The Philosophical Quarterly* 62 (247): 223–249.

Alfano, Mark. 2016. "The Topology of Communities of Trust." *Russian Sociological Review* 15 (4): 30–56.

Alfano, Mark, J. Adam Carter, and Marc Cheong. 2018. "Technological Seduction and Self-Radicalization." *Journal of the American Philosophical Association* 4 (3): 298–322.

Alfano, Mark, Amir Ebrahimi Fard, J. Adam Carter, Peter Clutton, and Colin Klein. 2021. "Technologically Scaffolded Atypical Cognition: The Case of YouTube's Recommender System." *Synthese* 199 (3–4): 835–858.

Allchin, Douglas. 2022. "Ten Competencies for the Science Misinformation Crisis." *Science Education* 107 (2): 261–274. https://doi.org.10.1002/sce.21746.

Allen-Hermanson, Sean. 2013. "Superdupersizing the Mind: Extended Cognition and the Persistence of Cognitive Bloat." *Philosophical Studies* 164 (3): 791–806. https://doi.org/10.1007/s11098-012-9914-7.

Altay, Sacha, and Alberto Acerbi. 2024. "People Believe Misinformation Is a Threat Because They Assume Others Are Gullible." *New Media & Society* 26 (11): 6440–6461. https://doi.org/10.1177/14614448231153379.

Altay, Sacha, and Mercier, Hugo. In press. Misinformation is a symptom: Commentary on Ecker et al. 2024. *American Psychologist.*

Altay, Sacha, Manon Berriche, Hendrik Heuer, Johan Farkas, and Steven Rathje. 2023. "A survey of expert views on misinformation: Definitions, determinants, solutions, and future of the field." *Harvard Kennedy School Misinformation Review* 4 (3): 1–34.

Alter, Adam. 2018. *Irresistible: The Rise of Addictive Technology and the Business of Keeping Us Hooked.* Penguin.

Anderson, Elizabeth. 2011. "Democracy, Public Policy, and Lay Assessments of Scientific Testimony." *Episteme* 8 (2): 144–164. https://doi.org/10.3366/epi.2011.0013.

Anderson, Joel, and Axel Honneth. 2005. "Autonomy, Vulnerability, Recognition, and Justice." In *Autonomy and the Challenges to Liberalism: New Essays*, edited by Joel Anderson and John Christman, 127–149. Cambridge University Press.

Anderson, Lisa R., and Charles A. Holt. 1997. "Information Cascades in the Laboratory." *American Economic Review* 87 (5): 847–862.

Andrada, Gloria. 2019. "Cognition as an Enculturated and Extended Social Skill." *Australasian Philosophical Review* (1): 71–75.

Arcuri, Maria Cristina, Gino Gandolfi, and Ivan Russo. 2023. "Does Fake News Impact Stock Returns? Evidence from US and EU Stock Markets." *Journal of Economics and Business* 125–126 (May–June): 106130. https://doi.org/10.1016/j.jeconbus.2023.106130.

Arvedlund, Erin. 2009. *Madoff: The Man Who Stole $65 Billion.* Penguin. https://books.google.co.uk/books?hl=en&lr=&id=spN9yCPn16MC&oi=fnd&pg=PT2&dq=madoff&ots=8sKGfDjVWC&sig=BxgAifbQImv4xAaBKxlXhzOiSww.

Audi, Robert. 1999. "Doxastic Voluntarism and the Ethics of Belief." *Facta Philosophica* 1 (1): 87–109. https://doi.org/10.5840/factaphil1999116.

Backes, Marvin. 2021. "Can Groups Be Genuine Believers? The Argument from Interpretationism." *Synthese* 199 (3–4): 10311–10329. https://doi.org/10.1007/s11229-021-03246-w.

Balakrishnan, Vimala, Wei Zhen Ng, Mun Chong Soo, Gan Joo Han, and Choon Jiat Lee. 2022. "Infodemic and Fake News—A Comprehensive Overview of Its Global Magnitude During the COVID-19 Pandemic in 2021: A Scoping Review." *International Journal of Disaster Risk Reduction* 78 (August): 103144. https://doi.org/10.1016/j.ijdrr.2022.103144.

Barberá, Pablo, John T. Jost, Jonathan Nagler, Joshua A. Tucker, and Richard Bonneau. 2015. "Tweeting from Left to Right: Is Online Political Communication More Than an Echo Chamber?" *Psychological Science* 26 (10): 1531–1542. https://doi.org/10.1177/0956797615594620.

Baribi-Bartov, Sahar, Briony Swire-Thompson, and Nir Grinberg. 2024. "Supersharers of Fake News on Twitter." *Science* 384 (6699): 979–982. https://doi.org/10.1126/science.adl4435.

Baron-Cohen, Simon. 1997. *Mindblindness: An Essay on Autism and Theory of Mind.* MIT Press.

Bawden, D., and Robinson, L. (2020, June 30). "Information Overload: An Introduction." *Oxford Research Encyclopedia of Politics*. Retrieved 20 Jan. 2026, from https://oxfordre.com/politics/view/10.1093/acrefore/9780190228637.001.0001/acrefore-9780190228637-e-1360.

Beer, Randall D. 1995. "A Dynamical Systems Perspective on Agent-Environment Interaction." *Artificial Intelligence* 72 (1-2): 173–215.

Begby, Endre. 2021. "Evidential Preemption." *Philosophy and Phenomenological Research* 102 (3): 515–530. https://doi.org/10.1111/phpr.12654.

Begby, Endre. 2024. "From Belief Polarization to Echo Chambers: A Rationalizing Account." *Episteme* 21 (2): 519–539. https://doi.org/10.1017/epi.2022.14.

Bendaña, Joseph, and Eric Mandelbaum. 2021. "The Fragmentation of Belief." In *The Fragmented Mind*, edited by Cristina Borgoni, Dirk Kindermann, and Andrea Onofri. Oxford University Press. https://doi.org/10.1093/oso/9780198850670.003.0004.

Bender, Emily M., Timnit Gebru, Angelina McMillan-Major, and Shmargaret Shmitchell. 2021. "On the Dangers of Stochastic Parrots: Can Language Models Be Too Big? 🦜." In *Proceedings of the 2021 ACM Conference on Fairness, Accountability, and Transparency (FAccT '21)*. Association for Computing Machinery. https://doi.org/10.1145/3442188.3445922.

Bengson, John, and Marc A. Moffett. 2011. "Two Conceptions of Mind and Action: Knowing How and the Philosophical Theory of Intelligence." In *Knowing How: Essays on Knowledge, Mind, and Action*, edited by John Bengson and Marc A. Moffett, 3–55. Oxford University Press. https://doi.org/10.1093/acprof:oso/9780195389364.003.0001.

Benkler, Yochai, Robert Faris, and Hal Roberts. 2018. *Network Propaganda: Manipulation, Disinformation, and Radicalization in American Politics*. Oxford University Press.

Benson, Paul. 2005. "Taking Ownership: Authority and Voice in Autonomous Agency." In *Autonomy and the Challenges to Liberalism: New Essays*, edited by John Christman and Joel Anderson, 101–126 (page range varies by edition). Cambridge University Press.

Benson, Paul. 1991. "Autonomy and Oppressive Socialization." *Social Theory and Practice* 17 (3): 385–408.

Bennett, Catherine. 2010. "Oh, Mr. Cameron, Do Stop all that Annoying Nudging." *The Guardian*. https://www.theguardian.com/commentisfree/2010/dec/05/nudge-health-policy-david-cameron.

Berinsky, Adam J. 2018. "Telling the Truth About Believing the Lies? Evidence for the Limited Prevalence of Expressive Survey Responding." *The Journal of Politics* 80 (1): 211–224. https://doi.org/10.1086/694258.

Bessi, Alessandro, Mauro Coletto, George Alexandru Davidescu, Antonio Scala, Guido Caldarelli, and Walter Quattrociocchi. 2015. "Science vs Conspiracy: Collective Narratives in the Age of Misinformation." *PLOS ONE* 10 (2): e0118093. https://doi.org/10.1371/journal.pone.0118093.

Bet, Marina. 2024. "Keir Starmer Unable to Define a Woman AGAIN as He Insists Trans Debate Must Be 'Treated with Respect.'" *Sun*, March 21. https://www.thesun.co.uk/news/26845883/keir-starmer-transgender-women-define-is/.

Bird, Alexander. 2010. "Social Knowing: The Social Sense of 'Scientific Knowledge.'" *Philosophical Perspectives* 24 (1): 23–56. https://doi.org/10.1111/j.1520-8583.2010.00184.x.

Bird, Alexander. 2014. "When Is There a Group That Knows?: Distributed Cognition, Scientific Knowledge, and the Social Epistemic Subject." In *Essays in Collective Epistemology*, edited by Jennifer Lackey. Oxford University Press.

Bird, Alexander. 2019. "Group Belief and Knowledge." In *The Routledge Handbook of Social Epistemology*, edited by Mirander Fricker, Peter J. Graham, David Henderson, and Nikolaj J. L. L. Pedersen. Routledge.

Bleakley, Paul. 2023. "Panic, Pizza and Mainstreaming the Alt-Right: A Social Media Analysis of Pizzagate and the Rise of the QAnon Conspiracy." *Current Sociology* 71 (3): 509–525. https://doi.org/10.1177/00113921211034896.

Blumenthal-Barby, J. S., and Hadley Burroughs. 2012. "Seeking Better Health Care Outcomes: The Ethics of Using the 'Nudge.'" *The American Journal of Bioethics* 12 (2): 1–10. https://doi.org/10.1080/15265161.2011.634481.

Bond, Charles, F., and Bella M. DePaulo. 2006. "Accuracy of Deception Judgments." *Personality and Social Psychology Review* 10 (3): 214–234.

Bortolotti, Lisa. 2022. "Delusion." In *The Stanford Encyclopedia of Philosophy*, edited by Edward N. Zalta. Metaphysics Research Lab, Stanford University. https://plato.stanford.edu/archives/sum2022/entries/delusion/.

Bostrom, Nick. 2003. "Are We Living in a Computer Simulation?" *Philosophical Quarterly* 53 (211): 243–255. https://doi.org/10.1111/1467-9213.00309.

Boult, Cameron. 2024. *Epistemic Blame: The Nature and Norms of Epistemic Relationships.* Oxford University Press.

Bovens, Luc. 2009. "The Ethics of Nudge." In *Preference Change: Approaches from Philosophy, Economics and Psychology*, edited by Till Grüne-Yanoff and Sven Ove Hansson. Theory and Decision Library. Springer. https://doi.org/10.1007/978-90-481-2593-7_10.

Boyd, Kenneth. 2019. "Epistemically Pernicious Groups and the Groupstrapping Problem." *Social Epistemology* 33 (1): 61–73. https://doi.org/10.1080/02691728.2018.1551436.

Boyd, Robert, Peter J. Richerson, and Joseph Henrich. 2011. "The Cultural Niche: Why Social Learning Is Essential for Human Adaptation." *Proceedings of the National Academy of Sciences of the United States of America* 108 (Supplement 2): 10918–10925. https://doi.org/10.1073/pnas.1100290108.

Brady, Michael S. 2016. *Emotional Insight: The Epistemic Role of Emotional Experience.* Oxford University Press.

Brancazio, Nick, and Neil Levy. 2024. "Do We Still Need Experts?" In *Philosophy, Expertise, and the Myth of Neutrality*, edited by Andrea Lavazza and Mirko Farina. Routledge. https://philpapers.org/rec/BRADWS-2.

Broadie, Sarah. 1991. *Ethics with Aristotle.* Oxford University Press.

Broncano-Berrocal, Fernando, and J. Adam Carter. 2021. *The Philosophy of Group Polarization: Epistemology, Metaphysics, Psychology*. Routledge.

Brooks, Patrick. 2023. "On the Origin of Conspiracy Theories." *Philosophical Studies* 180 (12): 3279–3299. https://doi.org/10.1007/s11098-023-02040-3.

Bullock, John G., and Gabriel Lenz. 2019. "Partisan Bias in Surveys." *Annual Review of Political Science* 22 (1): 325–342. https://doi.org/10.1146/annurev-polisci-051117-050904.

Caldwell, Noah. 2021. "America's Satanic Panic Returns—This Time Through QAnon." *NPR*, May 18, sec. National. https://www.npr.org/2021/05/18/997559036/americas-satanic-panic-returns-this-time-through-qanon.

Campbell, Douglas R. 2023. "In Defense of (Some) Online Echo Chambers." *Ethics and Information Technology* 25 (3): 45.

Carruthers, Peter. 2013. *The Opacity of Mind: An Integrative Theory of Self-Knowledge.* Oxford University Press.

Carter, J. Adam. 2022. *Autonomous Knowledge: Radical Enhancement, Autonomy, and the Future of Knowing.* Oxford University Press.

Carter, J. Adam. 2023. *Stratified Virtue Epistemology: A Defence.* Elements in Epistemology. Cambridge University Press. https://doi.org/10.1017/9781009067546.

Carter, J. Adam. 2024a. *A Telic Theory of Trust.* Oxford University Press.

Carter, J. Adam. 2024b. *Digital Knowledge: A Philosophical Investigation.* Taylor & Francis.

Carter, J. Adam, and Clayton Littlejohn. 2021. *This Is Epistemology: An Introduction.* John Wiley.

Cassam, Quassim. 2018. *Conspiracy Theories.* John Wiley & Sons.

Centers for Disease Control and Prevention. 2023. "Data Overview | Opioids | CDC." August 8. https://www.cdc.gov/opioids/data/index.html.

Cepollaro, Bianca, Maxime Lepoutre, and Robert Mark Simpson. 2023. "Counterspeech." *Philosophy Compass* 18 (1): e12890. https://doi.org/10.1111/phc3.12890.

Chait, Jonathan. 2024. "How J. D. Vance and the Online Right Sabotaged Trump at the Debate." *New York Magazine,* September 11. https://nymag.com/intelligencer/article/trump-debate-eating-pets-j-d-vance-cats-dogs-springfield.html.

Chargaff, Erwin, Rakoma Lipshitz, Charlotte Green, and M. E. Hodes. 1951. "The Composition of the Desoxyribonucleic Acid of Salmon Sperm." *Journal of Biological Chemistry* 192 (1): 223–230.

Cheng, Jun-Jun, Yun Liu, Bo Shen, and Wei-Guo Yuan. 2013. "An Epidemic Model of Rumor Diffusion in Online Social Networks." *The European Physical Journal B* 86 (1): 29. https://doi.org/10.1140/epjb/e2012-30483-5.

Christensen, David. 2007. "Epistemology of Disagreement: The Good News." *The Philosophical Review* 116 (2): 187–217. https://doi.org/10.1215/00318108-2006-035.

Christman, John Philip. 1989. *The Inner Citadel: Essays on Individual Autonomy.* Oxford University Press.

Christman, John. 2004. "Relational Autonomy, Liberal Individualism, and the Social Constitution of Selves." In *Relational Autonomy: Feminist Perspectives on Autonomy, Agency, and the Social Self,* edited by Catriona Mackenzie and Natalie Stoljar, 143–171. New York: Oxford University Press.

Christman, John. 2020. "Autonomy in Moral and Political Philosophy." In *The Stanford Encyclopedia of Philosophy*, edited by Edward N. Zalta. Metaphysics Research Lab, Stanford University. https://plato.stanford.edu/archives/fall2020/entries/autonomy-moral/.

Cinelli, Matteo, Gianmarco De Francisci Morales, Alessandro Galeazzi, Walter Quattrociocchi, and Michele Starnini. 2021. "The Echo Chamber Effect on Social Media." *Proceedings of the National Academy of Sciences of the United States of America* 118 (9): e2023301118. https://doi.org/10.1073/pnas.2023301118.

Clark, Andy. 2008. *Supersizing the Mind: Embodiment, Action, and Cognitive Extension*. Oxford University Press.

Clark, Andy. 2010. "Memento's Revenge: The Extended Mind, Extended." In *The Extended Mind*, edited by Richard Menary. MIT Press.

Clark, Andy. 2012. "Embodied, embedded, and extended cognition." In *The Cambridge Handbook of Cognitive Science*, edited by Keith Frankish and William M. Ramsey, 275. Cambridge University Press.

Clark, Andy. 2016. *Surfing Uncertainty: Prediction, Action, and the Embodied Mind*. Oxford University Press.

Clark, Andy, and David Chalmers. 1998. "The Extended Mind." *Analysis* 58 (1): 7–19.

Clark, Luke, and Michael Zack. 2023. "Engineered Highs: Reward Variability and Frequency as Potential Prerequisites of Behavioural Addiction." *Addictive Behaviors* 140: 107626. https://doi.org/10.1016/j.addictbeh.2023.107626.

Cohen, L. Jonathan. 1995. *An Essay on Belief and Acceptance*. Rev. ed. Oxford University Press.

Cohen, Marshall. 2024. "Elon Musk Peddles Debunked 2020 Election Conspiracies at First Solo Town Hall Supporting Trump | CNN Business." *CNN*, October 18. https://edition.cnn.com/2024/10/17/media/elon-musk-dominion-voting-misinformation/index.html.

Colburn, Ben. 2010. *Autonomy and Liberalism*. Routledge.

Confessore, Nicholas, and Karen Yourish. 2016. "$2 Billion Worth of Free Media for Donald Trump." *New York Times*, March 15. https://www.nytimes.com/2016/03/16/upshot/measuring-donald-trumps-mammoth-advantage-in-free-media.html.

Cookson, J. Anthony, Joseph E. Engelberg, and William Mullins. 2023. "Echo Chambers." *The Review of Financial Studies* 36 (2): 450–500. https://doi.org/10.1093/rfs/hhac058.

Coppock, Alexander. 2023. *Persuasion in Parallel: How Information Changes Minds About Politics*. Chicago Studies in American Politics. University of Chicago Press. https://press.uchicago.edu/ucp/books/book/chicago/P/bo181475008.html.

Dang, Haixin, and Liam Kofi Bright. 2021. "Scientific Conclusions Need Not Be Accurate, Justified, or Believed by Their Authors." *Synthese* 199: 8187–8203. https://doi.org/10.1007/s11229-021-03158-9.

Davidson, Donald 1970/2001. *Essays on Actions and Events* (2nd ed). Oxford University Press.

Del Vicario, Michela, Alessandro Bessi, Fabiana Zollo, et al. 2016. "The Spreading of Misinformation Online." *Proceedings of the National Academy of Sciences of the United States of America* 113 (3): 554–559. https://doi.org/10.1073/pnas.1517441113.

Dennett, Daniel C. 1975. "Brain Writing and Mind Reading." In *Minnesota Studies in the Philosophy of Science*, edited by K. Gunderson. University of Minnesota Press.

Dennett, Daniel C. 1989. *The Intentional Stance*. MIT Press.

Descartes, René. 2006. *A Discourse on the Method: Of Correctly Conducting One's Reason and Seeking Truth in the Sciences*. Oxford University Press.

Dicey Jennings, Carolyn. 2020. *The Attending Mind*. Cambridge University Press.

Dicey Jennings, Carolyn. 2022. *Attention and Mental Control*. Cambridge University Press. https://doi.org/10.1017/9781108982269.

Dogramaci, Sinan, and Miriam Schoenfield. 2025. "Why I Am Not a Boltzmann Brain." *Philosophical Review* 134 (1): 1–33. https://doi.org/10.1215/00318108-11591068.

Donegan, Moira. 2020. "QAnon Conspiracists Believe in a Vast Pedophile Ring. The Truth Is Sadder." *Guardian*, September 20, sec. Opinion. https://www.theguardian.com/commentisfree/2020/sep/20/qanon-conspiracy-child-abuse-truth-trump.

Doorn, Maarten van. 2023. "Advancing the Debate on the Consequences of Misinformation: Clarifying Why It's Not (Just) about False Beliefs." *Inquiry*, 1–27. https://doi.org/10.1080/0020174X.2023.2289137.

Douglas, Karen M., Robbie M. Sutton, and Aleksandra Cichocka. 2017. "The Psychology of Conspiracy Theories." *Current Directions in Psychological Science* 26 (6): 538–542. https://doi.org/10.1177/0963721417718261.

Duede, Eamon. 2022. "Instruments, Agents, and Artificial Intelligence: Novel Epistemic Categories of Reliability." *Synthese* 200 (6): 491. https://doi.org/10.1007/s11229-022-03975-6.

Duncker, Karl. 1945. "On problem-solving" (L. S. Lees, Trans.). *Psychological Monographs,* 58 (5), i–113.

Dworkin, Gerald. 1981. "The Concept of Autonomy." *Grazer Philosophische Studien* 12: 203–213.

Eisemann, Nora, Stefan Bunk, Trasias Mukama, Hannah Baltus, et al. 2025. "Nationwide Real-World Implementation of AI for Cancer Detection in Population-Based Mammography Screening." *Nature Medicine* 31 (3): 917–924.

Elga, Adam. 2007. "Reflection and Disagreement." *Noûs* 41: 478–502.

Elga, Adam. 2008. "Lucky to Be Rational." Paper presented on 6 June 2008 at the Bellingham Summer Philosophy Conference. https://www.princeton.edu/~adame/papers/bellingham-lucky.pdf.

Elzinga, Benjamin. 2022. "Echo Chambers and Audio Signal Processing." *Episteme* 19 (3): 373–393. https://doi.org/10.1017/epi.2020.33.

Emerson, Ralph Waldo. 1841. "Self-Reliance." In *Ralph Waldo Emerson, Essays*, edited by Ralph Waldo Emerson. James Munroe and Company.

Erduran, Sibel, and Olivia and Levrini. 2024. "The Impact of Artificial Intelligence on Scientific Practices: An Emergent Area of Research for Science Education." *International Journal of Science Education* 46 (18): 1982–1989. https://doi.org/10.1080/09500693.2024.2306604.

Evans, Gareth. 1982. *The Varieties of Reference*. 1st ed. Oxford University Press.

Evans-Pritchard, Edward Evan. 1976. *Witchcraft, Oracles, and Magic Among the Azande*. Oxford University Press.

Falbo, Arianna. 2021. "Inquiry and Confirmation." *Analysis* 81 (4): 622–631. https://doi.org/10.1093/analys/anab037.

Fan, Ruiping. 2015. "Informed Consent: Why Family-Oriented?" In *Family-Oriented Informed Consent: East Asian and American Perspectives*, edited by Ruiping Fan. Philosophy and Medicine. Springer International. https://doi.org/10.1007/978-3-319-12120-8_1.

Fantl, Jeremy. 2021. "Fake News vs. Echo Chambers." *Social Epistemology* 35 (6): 645–659. https://doi.org/10.1080/02691728.2021.1946201.

Feldman, Harriet, and Karl Friston. 2010. "Attention, Uncertainty, and Free-Energy." *Frontiers in Human Neuroscience* 4 (December): 215. https://doi.org/10.3389/fnhum.2010.00215.

Floridi, Luciano. 2014. *The Fourth Revolution: How the Infosphere Is Reshaping Human Reality*. Oxford: Oxford University Press.

Flynn, D. J., Brendan Nyhan, and Jason Reifler. 2017. "The Nature and Origins of Misperceptions: Understanding False and Unsupported Beliefs About Politics." *Political Psychology* 38 (S1): 127–150. https://doi.org/10.1111/pops.12394.

Frankfurt, Harry G. 1971. "Freedom of the Will and the Concept of a Person." *The Journal of Philosophy* 68 (1): 5–20. https://doi.org/10.2307/2024717.

Fraser, Rachel. 2023. "How to Talk Back: Hate Speech, Misinformation, and the Limits of Salience." *Politics, Philosophy & Economics* 22 (3): 315–335. https://doi.org/10.1177/1470594X231167593.

Fratantonio, Giada. 2024. "Can Epistemic Paternalistic Practice Make Us Better Epistemic Agents?" *Educational Theory* 74 (1): 108–122. https://doi.org/10.1111/edth.12623.

Fricker, Miranda. 2007. *Epistemic Injustice: Power and the Ethics of Knowing*. Oxford University Press.

Friedman, Jane. 2013. "Question-Directed Attitudes." *Philosophical Perspectives* 27 (1): 145–174. https://doi.org/10.1111/phpe.12026.

Friedman, Jane. 2020. "The Epistemic and the Zetetic." *Philosophical Review* 129 (4): 501–536.

Friedman, Marilyn. 2003. *Autonomy, Gender, Politics*. Oxford University Press.

Friston, Karl. 2009. "The Free-Energy Principle: A Rough Guide to the Brain?" *Trends in Cognitive Sciences* 13 (7): 293–301. https://doi.org/10.1016/j.tics.2009.04.005.

Friston, Karl. 2010. "The Free-Energy Principle: A Unified Brain Theory?" *Nature Reviews Neuroscience* 11 (2): 127–138. https://doi.org/10.1038/nrn2787.

Funkhouser, Eric. 2022. "A Tribal Mind: Beliefs That Signal Group Identity or Commitment." *Mind and Language* 37 (3): 444–464. https://doi.org/10.1111/mila.12326.

Furnham, Adrian, and Hua Chu Boo. 2011. "A literature review of the anchoring effect." *The Journal of Socio-Economics*, 40 (1): 35–42.

Ganapini, Marianna. 2023. "The Signaling Function of Sharing Fake Stories." *Mind and Language* 38 (1): 64–80.

Gao, Jian, and Dashun Wang. 2024. "Quantifying the Benefit of Artificial Intelligence for Scientific Research." *arXiv*. https://doi.org/10.48550/arXiv.2304.10578.

Gardiner, Georgi. 2022. "Attunement: On the Cognitive Virtues of Attention." In *Social Virtue Epistemology*, edited by Mark Alfano, Colin Klein, and Jeroen de Ridder. Routledge.

Garrett, R. Kelly. 2009. "Echo Chambers Online?: Politically Motivated Selective Exposure among Internet News Users." *Journal of Computer-Mediated Communication* 14 (2): 265–285.

Gelfert, Axel. 2018. "Fake News: A Definition." *Informal Logic* 38 (1): 84–117. https://doi.org/10.22329/il.v38i1.5068.

Gergely, György, Harold Bekkering, and Ildikó Király. 2002. "Rational Imitation in Preverbal Infants." *Nature* 415 (6873): 755–755. https://doi.org/10.1038/415755a.

Gerken, Mikkel. 2022. *Scientific Testimony: Its Roles in Science and Society*. Oxford University Press.

Germani, Federico, Andrew B. Pattison, and Monta Reinfelde. 2022. "WHO and Digital Agencies: How to Effectively Tackle COVID-19 Misinformation Online." *BMJ Global Health* 7 (8): e009483. https://doi.org/10.1136/bmjgh-2022-009483.

Gessen, Masha. 2011. *Perfect Rigour: A Genius and the Mathematical Breakthrough of a Lifetime*. Icon Books. https://books.google.com/books?hl=en&lr=&id=kNuywqWPxpQC&oi=fnd&pg=PT6&dq=Masha+lessen+perfect+rigor&ots=tC_Lonomqn&sig=35eem940LFe4si42RIaOfoR7PqM.

Gharpure, Radhika, Candis M. Hunter, Amy H. Schnall, et al. 2020. "Knowledge and Practices Regarding Safe Household Cleaning and Disinfection for COVID-19 Prevention—United States, May 2020." *MMWR. Morbidity and Mortality Weekly Report* 69 (23): 705–709. https://doi.org/10.15585/mmwr.mm6923e2.

Gilbert, Daniel T., Romin W. Tafarodi, and Patrick S. Malone. 1993. "You Can't Not Believe Everything You Read." *Journal of Personality and Social Psychology* 65 (2): 221–233.

Gilbert, Margaret. 1992. *On Social Facts*. Princeton University Press.

Goldberg, Sanford. 2013. "Epistemic Dependence in Testimonial Belief, in the Classroom and Beyond." *Journal of Philosophy of Education* 47 (2): 168–186.

Goldberg, Sanford C. 2011. "If That Were True I Would Have Heard About It by Now." *Social Epistemology: Essential Readings*, edited by A. Goldman and D. Whitcomb. Oxford University Press.

Goldenberg, Maya J. 2021. *Vaccine Hesitancy*. University of Pittsburgh Press.

Goldman, Alvin I. 2001. "Experts: Which Ones Should You Trust?" *Philosophy and Phenomenological Research* 63 (1): 85–110. https://doi.org/10.1111/j.1933-1592.2001.tb00093.x.

Graham, Matthew H. 2023. "Measuring Misperceptions?" *American Political Science Review* 117 (1): 80–102. https://doi.org/10.1017/S0003055422000387.

Graham, Peter. 2023. "What's Wrong with Testimony? Defending the Epistemic Analogy Between Testimony and Perception." In *The Oxford Handbook of Social Epistemology*, edited by Jennifer Lackey and Aidan McGlynn. Oxford University Press.

Greco, Daniel. 2014. "Could kk be ok?" *The Journal of Philosophy 111* (4):169–197.

Greco, John. 2020. *The Transmission of Knowledge*. Cambridge University Press.

Grundmann, Thomas. 2021. "The Possibility of Epistemic Nudging: Reply to My Critics." 10 (12): 28–35. https://philpapers.org/rec/GRUTPO-24.

Grundmann, Thomas. 2023. "The Possibility of Epistemic Nudging." *Social Epistemology* 37 (2): 208–218. https://doi.org/10.1080/02691728.2021.1945160.

Grundmann, Thomas. 2025. "Experts: What Are They and How Can Laypeople Identify Them?" In *Oxford Handbook of Social Epistemology*, edited by Jennifer Lackey and Aidan McGlynn. Oxford University Press.

Guerrero, Alexander. 2017. "Living with Ignorance in a World of Experts." In *Perspectives on Ignorance from Moral and Social Philosophy*, edited by Rik Peels. Routledge.

Habgood-Coote, Joshua. 2023. "Deepfakes and the Epistemic Apocalypse." *Synthese* 201 (3): 103.

Hadisi, Reza. 2023. "Autonomy as Practical Understanding." *Ergo: An Open Access Journal of Philosophy* 11: 34. https://philarchive.org/rec/HADAAP.

Haji, Ishtiyaque. 2008. *Incompatibilism's Allure: Principle Arguments for Incompatibilism*. Broadview Press.

Hakli, Raul. 2006. "Group Beliefs and the Distinction between Belief and Acceptance." *Cognitive Systems Research* 7 (2–3): 286–297. https://doi.org/10.1016/j.cogsys.2005.11.013.

Hannon, Michael. 2021. "Disagreement or Badmouthing? The Role of Expressive Discourse in Politics." In *Political Epistemology*, edited by Elizabeth Edenberg and Michael Hannon. Oxford University Press. https://doi.org/10.1093/oso/9780192893338.003.0017.

Hardwig, John. 1985. "Epistemic Dependence." *Journal of Philosophy* 82 (7): 335–349. https://doi.org/jphil198582747.

Hardwig, John. 1991. "The Role of Trust in Knowledge." *Journal of Philosophy* 88 (12): 693–708. https://doi.org/jphil199188121.

Harris, Keith Raymond. 2022. "Real Fakes: The Epistemology of Online Misinformation." *Philosophy & Technology* 35 (83): 1–24. https://doi.org/doi.org/10.1007/s13347-022-00581-9.

Harris, Keith Raymond. 2024. "The Simulation Argument Reconsidered." *Analysis* 84 (1): 23–31. https://doi.org/10.1093/analys/anad048.

Harris, Paul L. 2012. *Trusting What You're Told*. Harvard University Press.

Harris, Paul L., Melissa A. Koenig, Kathleen H. Corriveau, and Vikram K. Jaswal. 2018. "Cognitive Foundations of Learning from Testimony." *Annual Review of Psychology* 69 (1): 251–273. https://doi.org/10.1146/annurev-psych-122216-011710.

Harris, Tristan. 2019. "How Technology Is Hijacking Your Mind—from a Former Insider." *Thrive Global* (blog), October 16. https://medium.com/thrive-global/how-technology-hijacks-peoples-minds-from-a-magician-and-google-s-design-ethicist-56d62ef5edf3.

Hartwig, Maria, and Charles F. Bond. 2011. "Why do Lie-catchers Fail? A Lens Model Meta-analysis of Human Lie Judgments." *Psychological Bulletin* 137 (4): 643–659.

Haugeland, John. 1998. *Having Thought*. Harvard University Press. https://www.hup.harvard.edu/books/9780674004153.

Hausman, Daniel M., and Brynn Welch. 2010. "Debate: To Nudge or Not to Nudge." *Journal of Political Philosophy* 18 (1): 123–136.

Heersmink, Richard. 2015. "Dimensions of Integration in Embedded and Extended Cognitive Systems." *Phenomenology and the Cognitive Sciences* 14 (3): 577–598. https://doi.org/10.1007/s11097-014-9355-1.

Helton, Grace. 2018. "If You Can't Change What You Believe, You Don't Believe It." *Noûs* 54 (3): 501–526. https://doi.org/10.1111/nous.12265.

Helton, Grace. 2024. "On Being a Lonely Brain-in-a-Vat: Structuralism, Solipsism, and the Threat from External World Skepticism." *Analytic Philosophy* 65 (3): 353–373. https://doi.org/10.1111/phib.12291.

Henrich, Joseph. 2015. *The Secret of Our Success: How Culture Is Driving Human Evolution, Domesticating Our Species, and Making Us Smarter*. Princeton University Press.

Henrich, Joseph. 2020. *The Weirdest People in the World: How the West Became Psychologically Peculiar and Particularly Prosperous*. Penguin.

Henrich, Joseph, and Natalie Henrich. 2010. "The Evolution of Cultural Adaptations: Fijian Food Taboos Protect against Dangerous Marine Toxins." *Proceedings of the Royal Society B: Biological Sciences* 277 (1701): 3715–3724. https://doi.org/10.1098/rspb.2010.1191.

Henrich, Joseph, Steven J. Heine, and Ara Norenzayan. 2010. "The Weirdest People in the World?" *Behavioral and Brain Sciences* 33 (2–3): 61–83. https://doi.org/10.1017/S0140525X0999152X.

Heyes, Cecilia. 2014. "False Belief in Infancy: A Fresh Look." *Developmental Science* 17 (5): 647–659. https://doi.org/10.1111/desc.12148.

Hohwy, Jakob. 2013. *The Predictive Mind*. Oxford University Press.

Houlden, Shandell, Jaigris Hodson, George Veletsianos, Darren Reid, and Chris Thompson-Wagner. 2021. "The Health Belief Model: How Public Health Can Address the Misinformation Crisis beyond COVID-19." *Public Health in Practice* 2 (November): 100151. https://doi.org/10.1016/j.puhip.2021.100151.

Humphreys, Keith, Chelsea L. Shover, Christina M. Andrews, et al. 2022. "Responding to the Opioid Crisis in North America and beyond: Recommendations of the Stanford-Lancet Commission." *The Lancet* 399 (10324): 555–604. https://doi.org/10.1016/S0140-6736(21)02252-2.

Ibrahim, Nur. 2020. "Did Trump Fail to Denounce QAnon Conspiracy Theories?" Snopes. October 16. https://www.snopes.com//fact-check/trump-fail-to-denounce-qanon/.

Ichikawa, Jonathan. 2024. *Epistemic Courage*. Oxford University Press.

Ipsos. 2020. "More than 1 in 3 Americans Believe a 'Deep State' Is Working to Undermine Trump." Ipsos. December 30. https://www.ipsos.com/en-us/news-polls/npr-misinformation-123020.

Iyengar, Shanto, Gaurav Sood, and Yphtach Lelkes. 2012. "Affect, Not Ideology: A Social Identity Perspective on Polarization." *Public Opinion Quarterly* 76 (3): 405–431. https://doi.org/10.1093/poq/nfs038.

Jacobs, Phi. 2025. "Trump Officials Downplay Fake Citations in High-Profile Report on Children's Health." *Science,* May 30. https://www.science.org/content/article/trump-officials-downplay-fake-citations-high-profile-report-children-s-health.

Joly Chock, Valerie. 2020. "The Ethics and Applications of Nudges." *PANDION: The Osprey Journal of Research and Ideas* 1 (2): 5. https://digitalcommons.unf.edu/pandion_unf/vol1/iss2/5.

Joseph, Andrew. 2020. "CDC: Some Americans Are Misusing Cleaning Products—Including Drinking Them—in Effort to Kill Coronavirus." *STAT* (blog). June 5. https://www.statnews.com/2020/06/05/cdc-misusing-bleach-try-kill-coronavirus/.

Kahan, Dan. 2013. "Ideology, Motivated Reasoning, and Cognitive Reflection." *Judgment and Decision Making* 8 (4): 407–424.

Kant, Immanuel. 1991. "What is Enlightenment?" In *Political Writings*, 54–60. Cambridge University Press.

Katsafanas, Paul. 2019. "Fanaticism and Sacred Values." *Philosophers' Imprint* 19: 1–20.

Kelp, C. 2015. Understanding phenomena. *Synthese* 192 (12): 3799–3816.

Kelp, Christoph. 2020. "Theory of Inquiry." *Philosophy and Phenomenological Research* 103 (2): 359–384. https://doi.org/10.1111/phpr.12719.

Kelp, Christoph. 2021. *Inquiry, Knowledge, and Understanding*. Oxford University Press.

Kirchhoff, Michael David, and Will Newsome. 2012. "Distributed Cognitive Agency in Virtue Epistemology." *Philosophical Explorations* 15 (2): 165–180.

Knight, Dacre R.T., Christopher A. Aakre, Christopher V. Anstine, Bala Munipalli, Parisa Biazar, Ghada Mitri, Jose Raul Valery, Tara Brigham, Shehzad K. Niazi, Adam I.

Perlman, John D. Halamka, and Abd Moain Abu Dabrh. 2023. "Artificial Intelligence for Patient Scheduling in The Real-World Health Care Setting: A Metanarrative Review." *Health Policy and Technology* 12 (4): 100824.

Kokotajlo, Daniel, Scott Alexander, Thomas Larsen, Eli Lifland, and Romeo Dean. 2025. "AI 2027: A Scenario for Superhuman Artificial Intelligence." *AI Futures Project*. https://ai-2027.com/.

Kolodny, Andrew. 2020. "How FDA Failures Contributed to the Opioid Crisis." *AMA Journal of Ethics* 22 (8): 743–750. https://doi.org/10.1001/amajethics.2020.743.

Kuem, Jiyoung, Soumya Ray, Pei-Fang Hsu, and Lara Khansa. 2021. "Smartphone Addiction and Conflict: An Incentive-Sensitisation Perspective of Addiction for Information Systems." *European Journal of Information Systems* 30 (4): 403–424. https://doi.org/10.1080/0960085X.2020.1807101.

Kukla, Rebecca. 2012. ""Author TBD": Radical Collaboration in Contemporary Biomedical Research." *Philosophy of Science* 79 (5): 845–858. https://doi.org/10.1086/668042.

Kumar, K. P. Krishna, and G. Geethakumari. 2014. "Detecting Misinformation in Online Social Networks Using Cognitive Psychology." *Human-Centric Computing and Information Sciences* 4 (1): 14. https://doi.org/10.1186/s13673-014-0014-x.

Kumar, Victor. 2016. "Nudges and Bumps." *Georgetown Journal of Law & Public Policy* 14: 861.

Kunzru, Hari. 2020. "For the Lulz." *New York Review of Books*, March 26. https://www.nybooks.com/articles/2020/03/26/trolls-4chan-gamergate-lulz/.

Lackey, Jennifer. 2010. "A Justificationist View of Disagreement's Epistemic Significance." In *Social Epistemology*, edited by Adrian Haddock, Alan Millar, and Duncan Pritchard. Oxford University Press. https://www.oxfordscholarship.com/view/10.1093/acprof:oso/9780199577477.001.0001/acprof-9780199577477-chapter-15.

Lackey, Jennifer. 2018. "Group lies." *Lying: Language, Knowledge, Ethics, and Politics*: 262–284.

Lackey, Jennifer. 2021. *The Epistemology of Groups*. Oxford University Press.

Lavie, Nilli. "Attention, Distraction, and Cognitive Control Under Load." *Current Directions in Psychological Science* 19 (3): 143–148.

Lepoutre, Maxime. 2021. *Democratic Speech in Divided Times*. Oxford University Press.

Lepoutre, Maxime. 2024. "Narrative Counterspeech." *Political Studies* 72 (2): 570–589. https://doi.org/10.1177/00323217221129253.

Leviston, Zoe, Iain Walker, and Sarah Morwinski. 2013. "Your Opinion on Climate Change Might Not Be as Common as You Think." *Nature Climate Change* 3 (4): 334–337. https://doi.org/10.1038/nclimate1754.

Levy, Neil. 2018a. "Taking Responsibility for Health in an Epistemically Polluted Environment." *Theoretical Medicine and Bioethics* 39: 123–141.

Levy, Neil. 2018b. "You Meta Believe It." *European Journal of Philosophy* 26 (2): 814–826. https://doi.org/10.1111/ejop.12344.

Levy, Neil. 2019a. "Due Deference to Denialism: Explaining Ordinary People's Rejection of Established Scientific Findings." *Synthese* 196 (1): 313–327. https://doi.org/doi.org/10.1007/s11229-017-1477-x.

Levy, Neil. 2019b. "Nudge, Nudge, Wink, Wink: Nudging Is Giving Reasons." *Ergo, an Open Access Journal of Philosophy* 6 (10): 2019–2020. https://doi.org/10.3998/ergo.12405314.0006.010.

Levy, Neil. 2021. "Echoes of Covid Misinformation." *Philosophical Psychology* 36 (5): 931–948. https://doi.org/10.1080/09515089.2021.2009452.

Levy, Neil. 2022a. "Conspiracy Theories as Serious Play." *Philosophical Topics* 50 (2): 1–19.

Levy, Neil. 2022b. "Do Your Own Research!" *Synthese* 200 (5): 356. https://doi.org/10.1007/s11229-022-03793-w.

Levy, Neil. 2023. "Echoes of covid misinformation." *Philosophical Psychology* 36 (5): 931-948.

Levy, Neil. 2024a. "Believing in Shmeliefs." *Ergo, an Open Access Journal of Philosophy* 11 (18): 466-489. https://doi.org/10.3998/ergo.6158.

Levy, Neil. 2024b. "There's More Belief than Van Leeuwen Believes." *Mind & Language* 39 (4): 584–589.

Levy, Neil. 2025. "Betting on Scams." *Social Epistemology*, 1–13. https://doi.org/10.1080/02691728.2025.2527768

Levy, Neil, and Mark Alfano. 2019. "Knowledge from Vice: Deeply Social Epistemology." *Mind* 129 (515): 887–915. https://doi.org/10.1093/mind/fzz017.

Levy, Neil, and Russell Varley. 2024. "Mind the Guardrails: Epistemic Trespassing and Apt Deference." *Social Epistemology*, 1–17. https://doi.org/10.1080/02691728.2024.2400560.

Lewis, David. 1979. "Scorekeeping in a Language Game." *Journal of Philosophical Logic* 8 (3): 339.

Liddell, James. 2024. "Trump Drags New Animal into His Debunked Claims Haitian Migrants Are Eating Pets in Ohio." *Independent*, September 13. https://www.independent.co.uk/news/world/americas/us-politics/trump-eating-pets-migrants-ohio-geese-b2612259.html.

Litman, Leib, Zohn Rosen, Rachel Hartman, et al. 2023. "Did People Really Drink Bleach to Prevent COVID-19? A Guide for Protecting Survey Data against Problematic Respondents." *PLOS ONE* 18 (7): e0287837. https://doi.org/10.1371/journal.pone.0287837.

Lopez, Jesse, and D. Sunshine Hillygus. 2018. "Why So Serious?: Survey Trolls and Misinformation." SSRN Scholarly Paper ID 3131087. Social Science Research Network. https://doi.org/10.2139/ssrn.3131087.

Lyons, Derek E., Andrew G. Young, and Frank C. Keil. 2007. "The Hidden Structure of Overimitation." *Proceedings of the National Academy of Sciences of the United States of America* 104 (50): 19751–19756. https://doi.org/10.1073/pnas.0704452104.

Mackenzie, Catriona, and Natalie Stoljar, eds. 2000. *Relational Autonomy: Feminist Perspectives on Autonomy, Agency, and the Social Self.* Oxford University Press.

Mackenzie, Catriona. 2014. "The Importance of Relational Autonomy and Capabilities for an Ethics of Vulnerability." In *Vulnerability: New Essays in Ethics and Feminist Philosophy*, edited by Catriona Mackenzie, Wendy Rogers, and Susan Dodds, 33–59. Oxford University Press.

Mackenzie, Catriona, and Natalie Stoljar (eds.) 2000. *Relational Autonomy: Feminist Perspectives on Autonomy, Agency, and the Social Self.* Oxford University Press.

Maddox, Brenda. 2002. *Rosalind Franklin: The Dark Lady of DNA*. HarperCollins. https://bookreadfree.com/433908/10663447.

Magrath, Robert D., Tonya M. Haff, Pamela M. Fallow, and Andrew N. Radford. 2015. "Eavesdropping on Heterospecific Alarm Calls: From Mechanisms to Consequences." *Biological Reviews* 90 (2): 560–586. https://doi.org/10.1111/brv.12122.

Mallory, Fintan. 2023. "Fictionalism about Chatbots." *Ergo an Open Access Journal of Philosophy* 10: 38. https://doi.org/10.3998/ergo.4668.

Malmqvist, Lars. 2024. "Sycophancy in Large Language Models: Causes and Mitigations." arXiv:2411.15287 [cs.CL]. https://doi.org/10.48550/arXiv.2411.15287.

Mandelbaum, Eric. 2014. "Thinking Is Believing." *Inquiry: An Interdisciplinary Journal of Philosophy* 57 (1): 55–96. https://doi.org/10.1080/0020174X.2014.858417.

Marušić, Berislav, and Stephen J. White. 2023. "Disagreement and Alienation." *Philosophical Perspectives* 37 (1): 210–227. https://doi.org/10.1111/phpe.12197.

Mascaro, Olivier, and Dan Sperber. 2009. "The Moral, Epistemic, and Mindreading Components of Children's Vigilance towards Deception." *Cognition* 112 (3): 367–380. https://doi.org/10.1016/j.cognition.2009.05.012.

Matheson, Jonathan. 2024. "The Philosophy of Epistemic Autonomy: Introduction to Special Issue." *Social Epistemology* 38 (3): 267–273. https://doi.org/10.1080/02691728.2024.2335623.

Matheson, Jonathan, and Valerie Joly Chock. 2021. "The Possibility of Epistemic Nudging: Reply to Grundmann." *Social Epistemology Review and Reply Collective* 10 (8): 36–42.

McGowan, Mary Kate. 2003. "Conversational Exercitives and the Force of Pornography." *Philosophy and Public Affairs* 31 (2): 155–189. https://doi.org/10.1111/j.1088-4963.2003.00155.x.

McGowan, Mary Kate. 2004. "Conversational Exercitives: Something Else We Do with Our Words." *Linguistics and Philosophy* 27 (1): 93–111. https://doi.org/10.1023/b:ling.0000010803.47264.f0.

McGowan, Mary Kate. 2022. "On Salience and Sneakiness." In *Salience: A Philosophical Inquiry*, edited by Sophie Archer. Routledge.

McKenna, Michael. 2012. "Defending Nonhistorical Compatibilism: A Reply to Haji and Cuypers1." *Philosophical Issues* 22 (1): 264–280. https://doi.org/10.1111/j.1533-6077.2012.00229.x.

McKiernan, Jennifer, and Henry Zeffman. 2025. "Starmer Does Not Believe Trans Women Are Women, No 10 Says." *BBC News*, April 23. https://www.bbc.com/news/articles/crldey0z00ro.

Mele, Alfred R. 2001. *Autonomous Agents: From Self-Control to Autonomy*. Oxford University Press on Demand.

Mele, Alfred R. 2013. "Moral Responsibility, Manipulation, and Minutelings." *The Journal of Ethics* 17 (3): 153–166.

Menary, Richard. 2010. "Cognitive Integration and the Extended Mind." In *The Extended Mind*, edited by Richard Menary. MIT Press.

Menary, Richard. 2012. "Cognitive Practices and Cognitive Character." *Philosophical Explorations* 15 (2): 147–164.

Mercier, Hugo. 2020. *Not Born Yesterday: The Science of Who We Trust and What We Believe*. Princeton University Press.

Mercier, Hugo, Emmanuel Trouche, Hiroshi Yama, Christophe Heintz, and Vittorio Girotto. 2015. "Experts and Laymen Grossly Underestimate the Benefits of Argumentation for Reasoning." *Thinking & Reasoning* 21 (3): 341–355. https://doi.org/10.1080/13546783.2014.981582.

Merrefield, Clark. 2025. "AI and the News: What Researchers Learned from the AP+the BBC." *The Journalist's Resource*. March 4. https://journalistsresource.org/home/ai-ap-bbc/.

Metaxas, Panagiotis, and Samantha Finn. 2019. "Investigating the Infamous #Pizzagate Conspiracy Theory." *Technology Science*. https://techscience.org/a/2019121802/.

Michaels, Sean. 2023. "Why I Used AI to Write My Novel." *Macleans.Ca* (blog), October 4. https://macleans.ca/culture/ai-written-novel/.

Miller, Dale T., and Cathy McFarland. 1991. "When Social Comparison Goes Awry: The Case of Pluralistic Ignorance." In *Social Comparison: Contemporary Theory and Research*, edited by Jerry Suls and Thomas Ashby Wills. Lawrence Erlbaum Associates.

Millgram, Elijah. 2015. *The Great Endarkenment: Philosophy for an Age of Hyperspecialization*. Oxford University Press.

Mills, Charles W. 2007. "White Ignorance." In *Race and Epistemologies of Ignorance*, edited by Shannon Sullivan and Nancy Tuana. State University of New York Press.

Millson, Jared A. 2020. "Seeking Confirmation: A Puzzle for Norms of Inquiry." *Analysis* 80 (4): 683–693. https://doi.org/10.1093/analys/anaa017.

Mitchell, Peter, and Hazel Lacohée. 1991. "Children's Early Understanding of False Belief." *Cognition* 39 (2): 107–127. https://doi.org/10.1016/0010-0277(91)90040-B.

Miton, Helena, and Hugo Mercier. 2015. "Cognitive Obstacles to Pro-Vaccination Beliefs." *Trends in Cognitive Sciences* 19 (11): 633–636.

Miyazono, Kengo. 2023. "Epistemic Libertarian Paternalism." *Erkenntnis* 89: 3005–3024. https://doi.org/10.1007/s10670-023-00664-9.

Mole, Christopher. 2011. *Attention Is Cognitive Unison: An Essay in Philosophical Psychology*. Oxford University Press.

Moran, Paul. 2021. "The Ethics of Nudging the Planner." *Professional Planner*. https://www.professionalplanner.com.au/2021/08/the-ethics-of-nudging-the-client/.

Munton, Jessie. 2023. "Prejudice as the Misattribution of Salience." *Analytic Philosophy* 64 (1): 1–19. https://doi.org/10.1111/phib.12250.

Nagell, K., R. S. Olguin, and M. Tomasello. 1993. "Processes of Social Learning in the Tool Use of Chimpanzees (*Pan troglodytes*) and Human Children (*Homo sapiens*)." *Journal of Comparative Psychology* 107 (2): 174–186. https://doi.org/10.1037/0735-7036.107.2.174.

Nasar, Sylvia, and David Gruber. 2006. "Manifold Destiny." *New Yorker*, August 28. http://web2.mat.uniroma3.it/scuola_orientamento/echi_stampa/archivio/2006/stampa_dal_23-8-06/new-yorker-28-8-06.pdf.

Nedelsky, Jennifer. 2011. *Law's Relations: A Relational Theory of Self, Autonomy, and Law*. New York: Oxford University Press.

Nedelsky, Jennifer. 1989. "Reconceiving Autonomy: Sources, Thoughts and Possibilities." *Yale Journal of Law & Feminism* 1 (1): 7–36.

Nguyen, C. Thi. 2020. "Echo Chambers and Epistemic Bubbles." *Episteme* 17 (2): 141–161. https://doi.org/10.1017/epi.2018.32.

Nussbaum, Martha C. 2000. *Women and Human Development: The Capabilities Approach*. The Seeley Lectures. Cambridge University Press. https://doi.org/10.1017/CBO9780511841286.

O'Connor, Cailin, and James Owen Weatherall. 2019. *The Misinformation Age: How False Beliefs Spread*. 1st ed. Yale University Press.

Oreskes, Naomi, and Erik M. Conway. 2011. *Merchants of Doubt: How a Handful of Scientists Obscured the Truth on Issues from Tobacco Smoke to Global Warming*. A&C Black.

Oshana, Marina. 2006. *Personal Autonomy in Society*. Routledge.

Oshana, Marina A. L. 1998. "Personal Autonomy and Society." *Journal of Social Philosophy* 29 (1): 81–102.

Oswald, Frederick L., Gregory Mitchell, Hart Blanton, James Jaccard, and Philip E. Tetlock. 2013. "Predicting Ethnic and Racial Discrimination: A Meta-Analysis of IAT Criterion Studies." *Journal of Personality and Social Psychology* 105 (2): 171.

Palermos, S. Orestis. 2014. "Loops, Constitution, and Cognitive Extension." *Cognitive Systems Research* 27: 25–41.

Palermos, S. Orestis. 2015. "Active Externalism, Virtue Reliabilism and Scientific Knowledge." *Synthese* 192 (9): 2955–2986.

Palermos, S. Orestis. 2020. "Epistemic Collaborations: Distributed Cognition and Virtue Reliabilism." *Erkenntnis* 87: 1481–1500.

Palm, Viktoria, Tobias Norajitra, Oyunbileg von Stackelberg et al. 2022. "AI-Supported Comprehensive Detection and Quantification of Biomarkers of Subclinical Widespread Diseases at Chest CT for Preventive Medicine." *Healthcare* 10 (11): 2166.

Paris, Francesca, and Larry Buchanan. 2023. "35 Ways Real People Are Using A.I. Right Now." *New York Times*, April 14, sec. The Upshot. https://www.nytimes.com/interactive/2023/04/14/upshot/up-ai-uses.html.

Parmer, W. Jared. 2023. "Nudges, Nudging, and Self-Guidance Under the Influence." *Ergo* 9: 44. https://doi.org/10.3998/ergo.2913.

Pavese, Carlotta. 2022. "Practical Knowledge First." *Synthese* 200 (5): 375. https://doi.org/10.1007/s11229-022-03848-y.

Penna, Dominic. 2022. "Sir Keir Starmer Refuses to Answer When Asked Whether a Woman Can Have a Penis." *Telegraph*, March 28. https://www.telegraph.co.uk/politics/2022/03/28/keir-starmer-refuses-answer-whether-woman-can-have-penis-latest/.

Piedrahita, Oscar A., and Matthew Vermaire. 2025. "Nudging for Judging That p." *The Philosophical Quarterly*, pqaf026. https://doi.org/10.1093/pq/pqaf026.

Porot, Nicolas, and Eric Mandelbaum. 2021. "The Science of Belief: A Progress Report." *Wiley Interdisciplinary Reviews. Cognitive Science* 12 (2): e1539. https://doi.org/10.1002/wcs.1539.

Prior, Markus, Gaurav Sood, and Kabir Khanna. 2015. "You Cannot Be Serious: The Impact of Accuracy Incentives on Partisan Bias in Reports of Economic Perceptions." *Quarterly Journal of Political Science* 10 (4): 489–518. https://doi.org/10.1561/100.00014127.

Pritchard, Duncan. 2009. "The value of knowledge." *The Harvard Review of Philosophy* 16 (1): 86–103.

Pritchard, Duncan. 2014. "Knowledge and understanding." In *Virtue Epistemology Naturalized: Bridges Between Virtue Epistemology and Philosophy of Science,* edited by Abrol Fairweather, 315–327. Springer International Publishing.

Pritchard, Duncan. 2016. "Seeing It for Oneself: Perceptual Knowledge, Understanding, and Intellectual Autonomy." *Episteme* 13 (1): 29–42.

Pritchard, Duncan. 2020. "Anti-Risk Virtue Epistemology." In *Virtue-Theoretic Epistemology: New Methods and Approaches,* edited by J. Greco & C. Kelp, 203–224 (Cambridge UP, 2020).

Pröllochs, Nicolas, Dominik Bär, and Stefan Feuerriegel. 2021. "Emotions in Online Rumor Diffusion." *EPJ Data Science* 10 (1): 51. https://doi.org/10.1140/epjds/s13688-021-00307-5.

Pugh, Jonathan. 2020. "Four Distinctions Concerning Rationality." In *Autonomy, Rationality, and Contemporary Bioethics.* Oxford University Press. https://www.ncbi.nlm.nih.gov/books/NBK556863/.

Quine, W. V., and J. S. Ullian. 1978. *The Web of Belief.* 2nd ed. McGraw-Hill Education.

Quinton, Anthony. 1976. "Social Objects." *Proceedings of the Aristotelian Society* 76 (1): 1–27. https://doi.org/10.1093/aristotelian/76.1.1.

Rakow, Tim. 2022. "Adversarial Collaboration." In *Avoiding Questionable Research Practices in Applied Psychology,* edited by W. O'Donohue, A. Masuda, and S. Lilienfeld. Springer. https://doi.org/10.1007/978-3-031-04968-2_16.

Raz, Joseph. 1986. *The Morality of Freedom.* Clarendon Press.

Reid, Thomas. 1764/1997. *An Inquiry into the Human Mind on the Principles of Common Sense,* edited by Derek R. Brookes. Edinburgh University Press. https://campuspress.yale.edu/keithderose/files/2021/04/Reid-Inquiry-selections-21.pdf.

Reimann, Nicholas. 2020. "Some Americans Are Tragically Still Drinking Bleach as a Coronavirus 'Cure.'" *Forbes*, August 24. https://www.forbes.com/sites/nicholasreimann/2020/08/24/some-americans-are-tragically-still-drinking-bleach-as-a-coronavirus-cure/?sh=1add6ed36748.

Reuters. 2020. "Gargling with Bleach? Americans Misusing Disinfectants to Prevent Coronavirus, Survey Finds." *Reuters*, June 6, sec. World. https://www.reuters.com/article/idUSKBN23D06H/.

Richerson, Peter J., and Robert Boyd. 2008. *Not by Genes Alone: How Culture Transformed Human Evolution*. University of Chicago Press.

Ridder, Jeroen de. 2024. "What's So Bad About Misinformation?" *Inquiry* 67 (9): 2956–2978. https://doi.org/10.1080/0020174X.2021.2002187.

Rini, Regina. 2020. "Deepfakes and the Epistemic Backstop." *Philosophers' Imprint* 20 (24): 1–16.

Rini, Regina. 2021. "Weaponized Skepticism: An Analysis of Social Media Deception as Applied Political Epistemology." In *Political Epistemology*, edited by Elizabeth Edenberg and Michael Hannon. Oxford University Press. https://philarchive.org/rec/RINWSA.

Rini, Regina. 2023. "A Talking Cure for Autonomy Traps: How to Share Our Social World with Chatbots." https://philarchive.org/rec/RINATC.

Richter, Marjorie. 2025. "How AI is Transforming the Legal Profession." *Thomson Reuters Law Blog* 16 August. https://legal.thomsonreuters.com/blog/how-ai-is-transforming-the-legal-profession.

Ritchie, Stuart. 2020. *Science Fictions: Exposing Fraud, Bias, Negligence and Hype in Science*. Random House.

Robson, David. 2016. "Why Are People so Incredibly Gullible?' *BBC*, March 24. https://www.bbc.com/future/article/20160323-why-are-people-so-incredibly-gullible.

Roeber, Blake. 2024. *Political Humility: The Limits of Knowledge in Our Partisan Political Climate*. Routledge.

Rose, Joel. 2020. "Even If It's 'Bonkers,' Poll Finds Many Believe QAnon and Other Conspiracy Theories." *NPR*, December 30, sec. National. https://www.npr.org/2020/12/30/951095644/even-if-its-bonkers-poll-finds-many-believe-qanon-and-other-conspiracy-theories.

Ross, Don, Carla Sharp, Rudy E. Vuchinich, and David Spurrett. 2012. *Midbrain Mutiny: The Picoeconomics and Neuroeconomics of Disordered Gambling: Economic Theory and Cognitive Science*. MIT Press.

Ross, Robert M., and Neil Levy. 2023. "Expressive Responding in Support of Donald Trump: An Extended Replication of Schaffner and Luks (2018)." *Collabra: Psychology* 9 (1): 68054. https://doi.org/10.1525/collabra.68054.

Rupert, Robert D. 2004. "Challenges to the Hypothesis of Extended Cognition." *The Journal of Philosophy* 101 (8): 389–428.

Rupert, Robert D., and J. Adam Carter. 2026. *Epistemology in the Subpersonal Vale*. Oxford University Press.

Ryan, Shane. 2025. "Epistemic Environmentalism: Nudging and Epistemic Externalities." *Public Policy and Administration* 40 (3): 497–512. https://doi.org/10.1177/09520767251327733.

Saghai, Yashar. 2013. "Salvaging the Concept of Nudge." *Journal of Medical Ethics* 39 (8): 487–493.

Saint-Croix, Catharine. 2025. "The Epistemology of Attention." In *A Companion to Epistemology*, 3rd ed., edited by Kurt Sylvan, Ernest Sosa, Jonathan Dancy, and Matthias Steup. Wiley Blackwell.

Saul, Jennifer Mather. 2024. *Dogwhistles and Figleaves: How Manipulative Language Spreads Racism and Falsehood*. Oxford University Press.

Schaffner, Brian F., and Samantha Luks. 2018. "Misinformation or Expressive Responding? What an Inauguration Crowd Can Tell Us About the Source of Political Misinformation in Surveys." *Political Opinion Quarterly* 82 (1): 135–147.

Schmidt, Charles W. 2010. "A Closer Look at Climate Change Skepticism." *Environmental Health Perspectives* 118 (12): A536–A540. https://doi.org/10.1289/ehp.118-a536.

Schwengerer, Lukas. 2022. "Defending Joint Acceptance Accounts of Group Belief Against the Challenge from Group Lies." *Logos and Episteme* 13 (4): 421–428. https://doi.org/10.5840/logos-episteme202213435.

Schwitzgebel, Eric. 2002. "A Phenomenal, Dispositional Account of Belief." *Noûs* 36 (2): 249–275. https://doi.org/10.1111/1468-0068.00370.

Schwitzgebel, Eric. 2008. "The Unreliability of Naive Introspection." *Philosophical Review* 117 (2): 245–273.

Schwitzgebel, Eric. 2009. "Knowing Your Own Beliefs." *Canadian Journal of Philosophy* 39 (sup1): 41–62. https://doi.org/10.1080/00455091.2009.10717643.

Shah, Nishi. 2003. "How Truth Governs Belief." *Philosophical Review* 112 (4): 447–482.

Shah, Nishi, and J. David Velleman. 2005. "Doxastic Deliberation." *Philosophical Review* 114 (4): 497–534. https://doi.org/10.1215/00318108-114-4-497.

Shroff, Lilla. 2025. "The AI Takeover of Education Is Just Getting Started." *The Atlantic*, 12 August. https://www.theatlantic.com/technology/archive/2025/08/ai-takeover-education-chatgpt/683840/.

Shtulman, Andrew. 2006. "Qualitative Differences between Naïve and Scientific Theories of Evolution." *Cognitive Psychology* 52 (2): 170–194. https://doi.org/10.1016/j.cogpsych.2005.10.001.

Shtulman, Andrew, and Prassede Calabi. 2013. "Tuition vs. Intuition: Effects of Instruction on Naïve Theories of Evolution." *Merrill-Palmer Quarterly* 59 (2): 141–167. https://doi.org/10.1353/mpq.2013.0010.

Siegel, Susanna. 2017. *The Rationality of Perception*. Oxford University Press.

Simion, Mona. 2021a. *Shifty Speech and Independent Thought: Epistemic Normativity in Context*. Oxford University Press.

Simion, Mona. 2021b. "Testimonial contractarianism: A knowledge-first social epistemology." *Noûs* 55 (4): 891–916.

Simion, Mona. 2024a. "Knowledge and Disinformation." *Episteme* 21 (4): 1208–1219. https://doi.org/10.1017/epi.2023.25.

Simion, Mona. 2024b. "Disinformation, Politically Motivated Reasoning, and Knowledge Resistance." *European Review* 32 (4): 321–334. https://doi.org/10.1017/S1062798724000218.

Simion, Mona, and Christoph Kelp. 2025. "Information, Misinformation, Disinformation: A Unified Conceptual Framewo." In *Mis/Disinformation and Other Epistemic Pathologies*, edited by Mihaela Popa-Wyatt. Cambridge University Press.

Simpson, Robert Mark. 2013. "Un-Ringing the Bell: McGowan on Oppressive Speech and the Asymmetric Pliability of Conversations." *Australasian Journal of Philosophy* 91 (3): 555–575. https://doi.org/10.1080/00048402.2012.704053.

Singareddy, Sanjana, Vijay Prabhu SN, Arturo P Jaramillo et al. 2023. "Artificial Intelligence and Its Role in the Management of Chronic Medical Conditions: A Systematic Review." *Cureus* 15 (9): e46066. doi: 10.7759/cureus.46066.

Sosa, Ernest. 2007. *A Virtue Epistemology: Apt Belief and Reflective Knowledge, Volume I*. Oxford University Press.

Sosa, Ernest. 2009. *Reflective Knowledge: Apt Belief and Reflective Knowledge, Volume II*. Oxford University Press.

Sosa, Ernest. 2010. *Knowing Full Well*. Princeton University Press.

Sosa, Ernest. 2015. *Judgment and Agency*. Oxford University Press.

Sosa, Ernest. 2021. *Epistemic Explanations: A Theory of Telic Normativity, and What It Explains*. Oxford University Press.

Singer, Daniel J., Aaron Bramson, Patrick Grim, et al. 2019. "Rational Social and Political Polarization." *Philosophical Studies* 176 (9): 2243–2267. https://doi.org/10.1007/s11098-018-1124-5.

Sourati, Jamshid, and James A. Evans. 2023. "Accelerating Science with Human-Aware Artificial Intelligence." *Nature Human Behaviour* 7 (10): 1682–1696. https://doi.org/10.1038/s41562-023-01648-z.

Spelke, Elizabeth S., and Sanna Tsivkin. 2001. "Language and Number: A Bilingual Training Study." *Cognition* 78 (1): 45–88.

Sperber, Dan, Fabrice Clément, Christophe Heintz, et al. 2010. "Epistemic Vigilance." *Mind & Language* 25 (4): 359–393. https://doi.org/10.1111/j.1468-0017.2010.01394.x.

Sprevak, Mark. 2009. "Extended Cognition and Functionalism." *Journal of Philosophy* 106 (9): 503–527. https://doi.org/10.5840/jphil2009106937.

Stanovich, Keith E. 2018. "Miserliness in Human Cognition: The Interaction of Detection, Override and Mindware." *Thinking & Reasoning* 24 (4): 423–444.

Stelter, Brian, and Hadas Gold. 2025. "How Actual "Fake News" Caused the Market Whiplash." *CNN*, April 7. https://www.cnn.com/2025/04/07/media/fake-news-x-post-caused-market-whiplash/index.html.

Stelter, Brian, and Ken Olshansky. 2015. "How Much Does Donald Trump Dominate TV News Coverage? This Much." *CNN*, December 6. https://money.cnn.com/2015/12/06/media/donald-trump-nightly-news-coverage/.

Sterelny, Kim. 2012. *The Evolved Apprentice: How Evolution Made Humans Unique.* MIT Press.

Stoljar, Natalie. 2014. "Autonomy and Adaptive Preference Formation." In *Autonomy, Oppression, and Gender*, edited by Andrea Veltman and Mark Piper. Oxford University Press. https://doi.org/10.1093/acprof:oso/9780199969104.003.0011.

Stoljar, Natalie. 2000. "Autonomy and Feminist Intuitions." In *Relational Autonomy: Feminist Perspectives on Autonomy, Agency, and the Social Self,* edited by Catriona Mackenzie and Natalie Stoljar, 94–111. New York: Oxford University Press.

Stroud, Natalie Jomini. 2010. "Polarization and Partisan Selective Exposure." *Journal of Communication* 60 (3): 556–576. https://doi.org/10.1111/j.1460-2466.2010.01497.x.

Strawson, Peter. 1962. "Freedom and Resentment." *Proceedings of the British Academy* 48: 187–211. https://doi.org/10.1073/pnas.48.1.1.

Stubbersfield, Joe. 2023. "Conspiracy Theories: A Cultural Evolution Theory Approach." In *The Oxford Handbook of Cultural Evolution*, edited by Jamshid J. Tehrani, Jeremy Kendal, and Rachel Kendal. Oxford University Press. https://doi.org/10.1093/oxfordhb/9780198869252.013.72.

Sunstein, C. R. 2006. *Infotopia: How Many Minds Produce Knowledge*. Oxford University Press.

Sunstein, C. R. 2018. "Misconceptions About Nudges." *Journal of Behavioral Economics for Policy*, 2(1), 61–67.

Sunstein, Cass R. 2018. *Republic: Divided Democracy in the Age of Social Media*. Princeton, NJ: Princeton University Press.

Sunstein, Cass R., and Reid Hastie. 2008. "Four Failures of Deliberating Groups." *SSRN Electronic Journal*. https://doi.org/10.2139/ssrn.1121400.

Takayasu, Misako, Kazuya Sato, Yukie Sano, Kenta Yamada, Wataru Miura, and Hideki Takayasu. 2015. "Rumor Diffusion and Convergence during the 3.11 Earthquake: A Twitter Case Study." *PLoS One* 10 (4): e0121443.

Tennie, Claudio, Josep Call, and Michael Tomasello. 2009. "Ratcheting up the Ratchet: On the Evolution of Cumulative Culture." *Philosophical Transactions of the Royal Society B: Biological Sciences* 364 (1528): 2405–2415. https://doi.org/10.1098/rstb.2009.0052.

Thaler, Richard H., and Cass R. Sunstein. 2003. "Libertarian Paternalism." *The American Economic Review* 93 (2): 175–179.

Thaler, Richard H., and Cass R. Sunstein. 2008. *Nudge: Improving Decisions About Health, Wealth, and Happiness*. Yale University Press.

Tollefsen, Deborah. 2002. "Organizations as True Believers." *Journal of Social Philosophy* 33 (3): 395–410. https://doi.org/10.1111/0047-2786.00149.

Torcal, Mariano, Andres Reiljan, and Lisa Zanotti. 2023. "Editorial: Affective Polarization in Comparative Perspective." *Frontiers in Political Science* 5 (January). https://doi.org/10.3389/fpos.2023.1112238.

Tuomela, Raimo. 1992. "Group Beliefs." *Synthese* 91 (3): 285–318. https://doi.org/10.1007/bf00413570.

Tversky, Amos, and Daniel Kahneman. 1974. "Judgment under Uncertainty: Heuristics and Biases." *Science* 185 (4157): 1124–1131. https://doi.org/10.1126/science.185.4157.1124.

Uscinski, Joseph E. 2018. *Conspiracy Theories and the People Who Believe Them*. Oxford University Press.

Van der Linden, Sander. 2023. *Foolproof: Why We Fall for Misinformation and How to Build Immunity*. Fourth Estate.

Van Leeuwen, Neil. 2023. *Religion as Make-Believe: A Theory of Belief, Imagination, and Group Identity*. Harvard University Press.

Venkataramakrishnan, Siddharth, and Hannah Murphy. 2021. "Quitting QAnon: Why It Is So Difficult to Abandon a Conspiracy Theory." *Financial Times*, April 14. https://www.ft.com/content/5715176a-03b3-4ee9-a857-c50298ffe9da.

Vermaire, Matthew. 2022. "In Search of Doxastic Involuntarism." *Philosophical Studies* 179 (2): 615–631. https://doi.org/10.1007/s11098-021-01673-6.

Vernon, James. 2014. *Distant Strangers: How Britain Became Modern*. University of California Press.

Vitz, Rico. 2008. "Doxastic Voluntarism." In *Internet Encyclopedia of Philosophy*. https://philpapers.org/rec/UNKDV.

Vosoughi, Soroush, Deb Roy, and Sinan Aral. 2018. "The Spread of True and False News Online." *Science* 359 (6380): 1146–1151.

Walter, Jennifer, and Lovat T. Ross. 2014. "Relational Autonomy in Assisted Living: A Focus on Diverse Care Settings for Older Adults." *Journal of Aging Studies* 28: 1–9.

Watzl, Sebastian. 2011. "The Nature of Attention." *Philosophy Compass* 6 (11): 842–853. https://doi.org/10.1111/j.1747-9991.2011.00433.x.

Watzl, Sebastian. 2022. "The Ethics of Attention: An Argument and a Framework." In *Salience: A Philosophical Inquiry*, edited by Sophie Archer. Routledge.

Wellman, Henry M., David Cross, and Julanne Watson. 2001. "Meta-Analysis of Theory-of-Mind Development: The Truth about False Belief." *Child Development* 72 (3): 655–684. https://doi.org/10.1111/1467-8624.00304.

Westlund, Andrea C. 2009. "Rethinking Relational Autonomy." *Hypatia* 24 (4): 26–49.

Whitcomb, Dennis, Heather Battaly, Jason Baehr, and Daniel Howard-Snyder. 2017. "Intellectual Humility: Owning Our Limitations." *Philosophy and Phenomenological Research* 94 (3): 509–539.

Whiteley, Ella. 2022. "Harmful Salience Perspectives." In *Salience: A Philosophical Inquiry*, edited by Sophie Archer. Routledge.

Whiteley, Paul. 2022. "How Accurate Are the Polls When Forecasting Election Outcomes Well into the Future?—British Politics and Policy at LSE." *British Politics and Policy at LSE* (blog), July 12. https://blogs.lse.ac.uk/politicsandpolicy/how-accurate-are-polls-when-forecasting-election-outcomes/.

Wilkinson, T. M. 2013. "Nudging and Manipulation." *Political Studies* 61 (2): 341–355.

Willard-Kyle, Christopher. 2023. "The Knowledge Norm for Inquiry." *Journal of Philosophy* 120 (11): 615–640. https://doi.org/10.5840/jphil20231201134.

Williams, Bernard. 1973. "Deciding to Believe." In *Problems of the Self*. Cambridge University Press.

Williams, Daniel. 2023. "The Fake News about Fake News." *Boston Review*, June 7. https://www.bostonreview.net/articles/the-fake-news-about-fake-news/.

Williamson, Timothy. 2000. *Knowledge and its Limits*. Oxford University Press.

Wilson, Robert A., and Lucia Foglia. 2017. "Embodied Cognition." In *The Stanford Encyclopedia of Philosophy*, edited by Edward N. Zalta. Metaphysics Research Lab, Stanford University. https://plato.stanford.edu/archives/spr2017/entries/embodied-cognition/.

Winsberg, Eric. 2018. *Philosophy and Climate Science*. Cambridge University Press.

Winsberg, Eric, Bryce Huebner, and Rebecca Kukla. 2014. "Accountability and Values in Radically Collaborative Research." *Studies in History and Philosophy of Science Part A* 46: 16–23. https://doi.org/10.1016/j.shpsa.2013.11.007.

Wray, K. Brad. 2003. "What Really Divides Gilbert and the Rejectionists?" *ProtoSociology* 18–19 (July): 363–376. https://doi.org/10.5840/protosociology200318/1914.

Wu, Wayne. 2014. *Attention*. Routledge.

Zaller, John, and Stanley Feldman. 1992. "A Simple Theory of the Survey Response: Answering Questions Versus Revealing Preferences." *American Journal of Political Science* 36 (3): 579. https://doi.org/10.2307/2111583.

Index

Acceptance, joint, 42–43, 51–52, 59–60
Adaptive control, 130–134
Adaptive preference formation, 78
Adekola, A., 13
Agency. *See also* Agential ambiguity
  epistemic, 1–4
  extension via tools, 67, 75–76, 79
  loss of, 73–75
  opacity of contribution, 78–81
Agential ambiguity, 78–84
  definition of, 78–79
  ownership condition and, 77–79
  tools vs. collaborators, 65–66, 73–75, 79–84
AGI (artificial general intelligence), 65
Alfano, M., 22, 101, 104, 116, 123, 127
Anderson, E., ix, 16, 18, 19
Aristotle, vii, 98
Artificial intelligence (AI). *See also* Large language models
  epistemic autonomy and, 65–84, 93
  epistemic power and, 68
  epistemic risks of, 73–84
  responsibility and ownership, 73–83
  as scaffold, 66–72
Attention, 21–36
  autonomy and, 32–34
  belief vs., 24–28, 34
  capture, 24–30
  hijacking, 21–36, 101–102
Attractor-driven loops, 121–125
Autonomy. *See also* Epistemic autonomy
  moral vs. epistemic, 1–2, 61
  self-respect and, 1–2

Bayesian approaches, 23
Belief
  alienation from, 69–70
  vs. attention, 22–24, 28–34
  autonomy, relation to, 2–4
  constructed, 9–10
  false, 6
  of groups (*see* Group beliefs)
  instability of, 9–10
  self-attribution of, 7–11
  vs. "shmelief," 11, 50–53
  testimony and, 2–4
  unsheddable, 77–78
Bias, cognitive, 5–6, 105, 108–111, 113
Bird, A., 44, 48–50
Boyd, R., 14, 66
Broncano-Berrocal, F., 126
Burke and Wills expedition, 15–16

Campbell, 124, 129
Carruthers, P., 11
Carter, A., 62, 93, 126, 138
Chalmers, D., 129
Choice architecture, 103–108, 113, 116–117
Christensen, D., 68–69

Clark, A., 95, 128–129
Cognitive tools
vs. collaborators, 74–75
integration into agency, 75–76
LLMs as, 68–69
Collaboration, 66–67
with AI, 73–75, 83–84
epistemic limits of, 82–83
epistemic value and, 90–95
with humans, 82–84
Collective rationality, 45–50
Conspiracy theories, 6–7,11, 22–23, 51–52, 103, 114, 121–122, 124, 133, 134
Continuous reciprocal causation (CRC), 128
Control theory, 131
Credibility
assessment of, 12–13
epistemic vigilance and, 12–14, 16, 28
Credit, epistemic, 75, 80, 82–83
Cultural evolution, 14–16
epistemic dependence and, 14–15, 66
imitation in, 15–16
Curiosity, 104–106

Data processing, 68
Deception
detection of, 11–13
Decision procedures, 40–44
Deference, epistemic, 3, 54, 86–88
Dennett, D., 41
Descartes, René, vii–ix
Disagreement
with LLMs, 71–72
peer, 69–72
Distributed models of group belief, 42–50, 52–53, 61–62
Division of labor, epistemic, 14, 49, 66, 120
Doxastic nudging, 103–104
vs. erotetic nudging, 104–106
Doxastic voluntarism, 24, 107

Echo chambers, 121–124, 129–130, 134
Elga, A., 68
Emerson, Ralph Waldo, 2–3
Enlightenment, 3
Epistemic autonomy, 1–5, 65–84, 85–102, 119–136
AI and, 65–84
dependence and, 1–5, 14–18
development of, 70–72
group autonomy, 37–64
negative autonomy, 85–102
ownership condition, 78–81
rationality and, 5–6
threats to, 73–84, 120–125
Epistemic dependence, 1–18, 90–102
cultural evolution and, 14–15
expertise and, 16–18
synchronic vs. diachronic, 94–95
testimony and, 2–4
Epistemic environments
high-density, 13
inquiry and, 105–110
loop-structured, 120–125
Epistemic goods
cognitive achievement, 86–90
understanding, 86–89
Epistemic power, 68
Epistemic scaffolding, 66–67
Epistemic vigilance, 12–14
limits of, 13–14
mechanisms of, 12–13
Equal-weight view, 68–69
Erotetic autonomy, 104–110
Erotetic nudging, 104–112
vs. manipulation, 110–112
types of, 106–110
Evidential environment, 107
Expertise
identification of, 16–17
limits of verification, 16–18

Experts
disagreement among, 16–18
Expressive responding, 7–9
Extended mind, 40, 128–129

Fake news, 5–6, 8, 23
Feedback loops, epistemic, 119–125
beneficial vs. harmful, 121–125
performative loops, 120–125
Firsthand knowledge (Sosa), 88
Flynn, D., 6
Framing effects, 106–110
Frankfurt, H., 60, 62, 131
Friedman, J., 103–104, 107
Friedman, M., 1

Gelfert, A., 5
Generative loops, 127129
Gilbert, M., 42
Goldberg, S., 32, 127
Goldman, A., ix, 16, 18
Group belief, 37–64
autonomy of, 37–64
non-summativism, 38–41
Group doxastic autonomy, 37–64
Group-level properties
emergence of, 38–41
irreducibility of, 41–44
Group reasoning
aggregation of judgments, 45–48
structure of, 40–44
Groupstrapping, 126–127
Groupthink, 53–55, 62
Grundmann, T., 16, 103, 106–107, 114
Guerrero, A., ix, 16
Gullibility, x, 5–6, 11, 22, 28

Hadisi, R., 131–132
Harris, K. R., 20
Harris, P. L., 12, 16
Haugeland, J., 67
Henrich, J., 2, 14–16, 66
Heteronomy, 59–62
Hijacked attention, 21–36
Humphreys, K., ix

Independence, epistemic, 3, 85–92, 94–97
critique of, 85–92
instrumental value of, 90–95
Information environment, 13
Inquiry
shaping of, 105–110
structure of, 104–106
Integration
of cognitive resources, 95
of groups, 49, 62
of tools into agency, 75–76
Intellectual autonomy. *See* Epistemic autonomy

Kant, Immanuel, 3–4
Kantianism, 73–74
Knowledge
distributed production of, 82–83
testimony and, 2–4

Lackey, J., 10, 43, 70
Large language models (LLMs). *See also* Artificial intelligence
as cognitive tools, 68–69
as collaborators, 73–75
epistemic autonomy and, 65–84
epistemic risks of, 73–84
as scaffolding, 66–67
self-trust and, 70–72
Learning, 70–72
Levy, Neil, ix, 7, 10–12, 16–19, 22–23, 41, 51, 104, 115
Loop-ridden landscape, 119–136
Luks, Samantha, 7–8

Manipulation, 110–112
Marlboro Man model, 1–2, 85–86
Marušić, B., 69

Millgram, E.,17, 66
Misinformation, x, 5–13, 22–28, 35, 68, 104, 118, 124
  attention-based harms, 30–34
  belief formation and, 6–7
  expressive responding, 7–8
  trolling, 8–9
Moral autonomy vs. epistemic autonomy, 1–2, 61

Negative autonomy thesis (NEA), 85–102
  critique of, 86–95
Nguyen, C. T., 121 133
Non-summativism, 38–41
Novice/expert problem, 16–19
Nudging, epistemic, 103–112
  doxastic vs. erotetic, 103–106

Ofcom report, 3
Opioid epidemic, viii–ix
Ownership condition, 78–81

Peer disagreement
  and autonomy, 69–72
Plausibility, 14–15
Polarization, group, 125–127
Ponzi scheme (Madoff), 127–128
Practical understanding, 131–132
Predictive coding, 24
Pritchard, Duncan, xi, 14, 87–92, 100
Problematic questions, 104–106

QAnon, 6–7, 22, 27
Quasi-testimony, 147n8

Ratchet effect, 66
Rationality, 5–6
Reductio argument, 94–95
Reid, T., 85
Repeller-driven loops, 125–127
Reputation, 12–13
Responsibility, epistemic, 79–83
Richerson, P., 66
Rini, A., 73–75
Risk opacity, 130–131
Robson, D., 22

Saghai, Y., 105–106, 114
Schaffner, Brian, 7–8
Schwitzgebel, Eric, 10–11, 41
Scientific research groups, 44–47, 66
Self-knowledge, 11
Self-regulation, 125–132
Self-trust, 71–72
Shmelief, 11, 41
  group, 50–52
Simion, Mona, 6, 12–13
Skepticism, Cartesian, vii–viii
Slot machine analogy, 73–75, 78
Social epistemology, 5, 119
Sosa, E., 87–89, 91–92
Sperber, Dan, 12–13, 16, 29
Star networks, 127–128
Summativism, 38–41
Sunstein, C., 54, 103, 105, 114, 133

Telic virtue epistemology, 88
Testimony, 2–4, 6, 12–18, 22–23, 53, 65, 82, 89, 91–93, 95–96, 101, 126, 128
Thaler, R., 103, 105, 114
Tools
  vs. collaborators, 74–75
  epistemic, 67
Trolling, 8–9
Trust
  experts and, 16–20
Tuomela, R., 42

Uncertainty
  of AI status, 74–75
Understanding, practical, 130–134
Understanding-why, 86–89

Vaccine hesitancy, 13
Van Leeuwen, N., 38–42

Vigilance, epistemic. *See* Epistemic vigilance
Virtue epistemology, 88
Vulnerability, epistemic, 14

Watzl, S., 21
White, R., 68–71
Winsberg, Eric, 17, 82
Wonder, 104, 107–108

Zetetic autonomy. *See* Erotetic autonomy
Zetetic turn, 86, 103–106

Publisher contact:
The MIT Press
Massachusetts Institute of Technology
77 Massachusetts Avenue, Cambridge, MA 02139
mitpress.mit.edu

EU Authorised Representative:
Easy Access System Europe, Mustamäe tee 50,
10621 Tallinn, Estonia
gpsr.requests@easproject.com

Printed by Integrated Books International,
United States of America